Parallel Processing

The Transputer and its Applications

INTERNATIONAL COMPUTER SCIENCE SERIES

Consulting editor **A D McGettrick** University of Strathclyde

SELECTED TITLES IN THE SERIES

Parallel Programming *R H Perrott*

Concurrent Programming *N Gehani and A D McGettrick*

Program Derivation: The Development of Programs from Specifications *R G Dromey*

Object-Oriented Programming with Simula *B Kirkerud*

Parallel Processing: Principles and Practice *E V Krishnamurthy*

Real Time Systems and Their Programming Languages *A Burns and A Wellings*

Fortran 77 Programming *T M R Ellis*

Prolog Programming for Artificial Intelligence (2nd Edn) *I Bratko*

Introduction to Expert Systems (2nd Edn) *P Jackson*

Computer Architecture *M De Blasi*

The Programming Process: an Introduction using VDM and Pascal *J T Latham, V J Bush and D Cottam*

Analysis of Algorithms and Data Structures *L Banachowski, A Kreczmar and W Rytter*

Handbook of Algorithms and Data Structures in Pascal and C (2nd Edn) *G Gonnet and R Baeza-Yates*

Algorithms and Data Structures *J H Kingston*

Principles of Expert Systems *P Lucas and L van der Gaag*

Discrete Mathematics for Computer Scientists *J K Truss*

Programming in Ada plus an Overview of Ada 9X (4th Edn) *J G P Barnes*

Software Engineering (4th Edn) *I Sommerville*

Distributed Database Systems *D Bell and J Grimson*

Software Development with Z *J B Wordsworth*

Program Verification *N Francéz*

Concurrent Systems: an Integrated Approach to Operating Systems, Database and Distribution Systems *J Bacon*

Comparative Programming Languages (2nd Edn) *L B Wilson and R G Clark*

Programming Language Essentials *H E Bal and D Grune*

Distributed Systems: Concepts and Design (2nd Edn) *G Coulouris, J Dollimore and T Kindberg*

Human–Computer Interaction *J Preece et al.*

Fortran 90 Programming *T M R Ellis, I R Philips and T M Lahey*

Parallel Processing

The Transputer and its Applications

Edited by

M.E.C. Hull

The University of Ulster

D. Crookes

The Queen's University, Belfast

P.J. Sweeney

The University of Ulster

ADDISON-WESLEY
PUBLISHING
COMPANY

Wokingham, England • Reading, Massachusetts • Menlo Park, California • New York
Don Mills, Ontario • Amsterdam • Bonn • Sydney • Singapore
Tokyo • Madrid • San Juan • Milan • Paris • Mexico City • Seoul • Taipei

Cover designed by Chris Eley
and printed by The Riverside Printing Co. (Reading) Ltd.
Typeset by PanTek Arts, Maidstone, Kent.
Printed in Great Britain at the University Press, Cambridge

First printed 1994.

ISBN 0-201-62755-8

British Library Cataloguing in Publication Data
A catalogue record for this book is available from the British Library.

Library of Congress Cataloging in Publication Data
Parallel processing : the transputer and its applications / edited by
M.E.C. Hull, D. Crookes, P.J. Sweeney.
p. cm. -- (International computer science series)
Includes bibliographical references and index.
ISBN 0-201-62755-8
1. Parallel processing (Electronic computers) 2. Transputers.
I. Hull, M.E.C. II. Crookes, Daniel, date. III. Sweeney, P.
J. (Patrick J.) IV. Series.
QA76.58.P3783 1994
004. 3'5 --dc20 94-14083
CIP

Preface

The motivation for this book arose from courses given at the Northern Ireland Regional Transputer Support Centre, where I was the Director for four years. It is concerned not only with the transputer, its architecture and the programming language occam, but, since the Transputer Centre considered technology transfer as one of its main aims, the book identifies a variety of application areas.

The book is targeted at undergraduate and postgraduate students of computer science, electronic engineering and mathematics. It will also be suitable for project managers and software engineers in industry who require a text which combines details of the transputer with more general issues relating to parallel processing applications.

The book falls neatly into three main parts. The first three chapters concentrate on the underlying issues relating to parallel processing, and in particular the development of the programming language occam plus other languages which are used to program the transputer.

Chapter 2 provides an introductory course on the programming language occam, and no prior knowledge is required. The chapter is illustrated with numerous examples to assist the reader in the learning process. Chapter 3 addresses those in industry who, while willing to use parallel architectures, wish to use much more conventional programming languages. In response to this, software houses have developed compilers for parallel versions of familiar languages such as C, Fortran and Pascal. This chapter shows the reader how a sequential program in one of these languages can be readily run on a single transputer, how the program might be parallelized to run as several processes on a single transputer, and how it might then be run on a network of transputers to achieve true concurrency.

Part II of the book concentrates on the hardware used in parallel processing and the context within which transputer systems have developed. Chapter 4 provides a brief overview of low-level transputer hardware, together with the software origins of the transputer; additionally, some building blocks and interfacing components are introduced. Chapter 5 then describes many of the available board and system level components which can be integrated to make systems of various sizes.

Part III attempts to answer the question 'Are transputers useful in practice, and can they be used to construct real applications?' The chapters are a collection of case studies from a wide range of application areas which are being studied in practice, and they provide a valuable insight to both academics and industrialists seeking to identify potential application areas for parallel processing.

Acknowledgements

I am grateful to The University of Ulster and The Queen's University of Belfast for their cooperation and support during the lifetime of the Northern Ireland Regional Transputer Support Centre. I am also indebted to the Science and Engineering Research Council and the Department of Economic Development (Northern Ireland) for funding the Transputer Centre. I would like to record my thanks to the academic, technical and secretarial staff associated with the Centre, without whose help and support the operation of the Transputer Centre would not have been possible, and neither would this text. I am particularly grateful to the authors of the individual chapters.

A special word of thanks is due to Paul Stewart of the University of Ulster for all his help in presenting the text, which he received in various styles from the authors, in a suitable format acceptable to the publishers.

Finally, I would like to express my thanks to my co-editors, D. Crookes and P. J. Sweeney, together with the series editor A. D. McGettrick and the publishers.

M. E. C. HULL
June 1994

Contributors

K. Adamson, M.E.C. Hull, R.J. Millar, P.J. Morrow, J. Shao and P.J. Sweeney are at the Department of Computing Science, The University of Ulster.

W.A. Coey, D. Crookes, J.W. Flanigan, D.R.W. Holton, R.M. McKeag, J.D.M. McKeever and J.R. Stewart are at the Department of Computer Science, The Queen's University of Belfast.

Contents

Part I

SOFTWARE FOR PARALLELISM

Chapter 1

Introduction

M.E.C. Hull

1.1 Setting the scene

The concept of developing parallel, or concurrent, programs is not a recently introduced idea. This is substantiated if we look at work which proposed the use of parallelism in the development of system software in the 1960s, and by looking at the range of notations proposed over the past 20 years such as Communicating Sequential Processes (CSP) and programming languages such as Pascal Plus, Ada and occam, which were developed within the past decade specifically for the purpose of expressing parallelism in programs. What do we mean by 'parallel programs' and 'parallel programming'?

Within the context of this book, we define a **parallel program** as a program composed of a number of sections of code where all the sections can be executed at the same time: it is assumed that each section of code will be resident on its own processor. This is a very primitive definition, and we will have to consider its implications very carefully to determine what type of hardware and software will be required to permit the successful development of programs of this type.

Although there are several different forms, or types, of parallelism, a considerable portion of the research and development work in the field of parallel programming has been concentrated in two areas:

(1) SIMD parallelism.

(2) MIMD parallelism.

SIMD, or Single Instruction Multiple Data, parallelism lies within the realm of the so-called 'supercomputers' such as the CRAY and the DAP. As its name suggests, this type of parallelism offers a world in which the same instruction can be applied to a group of related data items within the same time slice. To illustrate this, consider the following trivial statement which is representative of programming language instructions with the SIMD environment:

$$A(*) = B(*) + C(*)$$

where A, B and C are matrices and the '*' is used to note that a parallel operation is to be performed. This statement implies that the following set of instructions will be executed concurrently:

$$\begin{aligned} A(1) &= B(1) + C(1) \\ A(2) &= B(2) + C(2) \\ &\vdots \\ A(n) &= B(n) + C(n) \end{aligned}$$

The range over which the parallel operation will be applied is usually fixed, for example the CRAY is a **vector processor** so the operations will be applied to a one-dimensional matrix of length 64, whereas the DAP is an **array processor** where the operations will be applied to a two-dimensional matrix of dimension 64 × 64. It should be noted that this is an overly-simplified picture of SIMD parallelism, but it will suffice for setting the scene. In the example given above, note the *level* at which the parallelism is represented. In this case, it is the statement level, that is within a single statement parallel activity is represented explicitly. This can be regarded as **fine grain parallelism**.

MIMD, or Multiple Instruction Multiple Data, parallelism offers a completely different approach to parallel programming. Within the MIMD environment it is usual to find that groups of (unrelated) statements are executed in parallel. To illustrate this consider the following simple example based on the structure of a typical compiler. Each of the main sections of the compiler will consist of a large number of statements, but it is possible to ignore this fine detail and regard each of the sections as an independent block (entity, process, task, object,...) and demand that all such blocks be executed in parallel. For example, in Hoare's CSP notation this could be expressed as follows

[source handler // lexical analyser //.... // code generator]

where the local case identifiers represent the processes and the parallel bars are used to denote that the processes are to be executed in parallel.

Note that the level (or degree) of parallelism in an MIMD system differs from that present in an SIMD environment – in an MIMD environment a program can be said to exhibit **course grain parallelism**, and it is on parallelism of this type that we shall concentrate in this book.

There are two main groups of users interested in parallelism and this form of architecture:

(1) Users who require considerable computational power – processing rates well beyond the capability of traditional von Neumann architectures. This group includes physicists, applied mathematicians, engineers, etc., and they regard parallel systems as a means of achieving considerable reductions in program execution times.

It is fair to say that this group of users is not interested in the architectures and programming languages used in parallel programming. They confine themselves, in the main, to the use of array and vector processors where the most popular languages are 'parallel' versions of Fortran.

(2) The second group can be viewed as containing those people concerned with the problems associated with the development of

(a) programming methodologies to assist with the production of parallel programs in the same way that methodologies such as stepwise refinement were developed to assist with the production of sequential software;

(b) programming languages with the necessary structures for representing parallelism;

(c) formal techniques for the specification and proof of parallel programs.

Not surprisingly, this group tends to be composed, for the most part, of computer scientists.

In spite of the interest in the development of MIMD programs and the very rapid development of suitable parallel programming languages, the lack of commercially available MIMD hardware has imposed a severe restriction on the acquisition of real experience in the implementation of parallel programs.

The advent of the transputer in October 1985 heralded a major change in this area as it became possible to construct flexible multiprocessor systems with ease. But

What is a **transputer**?

What do we mean by **flexible multiprocessor**?

A typical member of the transputer family can be viewed as a single chip that contains a processor, memory and communication links (which provide point-to-point connection between transputers). In other words, a transputer can be thought of as providing a complete von Neumann system on a single chip.

In this book, the target architecture will be regarded as being a multiprocessor which can be defined as a system composed as a number of possibly different processors, where each processor has its own memory and has the ability to communicate with any other processor in the system. The term 'flexible multiprocessor' can be explained by examining the possible forms of processor topology offered by the current range of MIMD systems.

There are two features of MIMD architectures that can be used to classify such systems:

(1) The number of processors in the system.

(2) The manner in which the processors are interconnected.

In general, three types of system can be identified:

(1) A system where both the number of processors in the system and the way in which they are interconnected are fixed. For example, the Intel iPSC and the FPS T-series.

(2) A system where the number of processors in the system can be varied and the processor topology can be reconfigured, that is a flexible multiprocessor system; for example, a system composed of Inmos B002 or B004 transputer development boards.

(3) A system which falls midway between types (1) and (2), that is a system which exhibits some degree of rigidity; for example, a system composed of Inmos B003 transputer development boards.

The two features used to classify MIMD systems are of considerable importance as they have a direct effect on the efficiency of a parallel program.

If the number of processes in an algorithm exceeds the number of processors in the system, then some **multi-tasking** is inevitable, and unless the **load** on the processors is approximately the same bottlenecks will occur. Hence there is a need to conduct load balancing studies to ensure that the effect of multi-tasking is minimal.

The manner in which the processors are interconnected is important, as the use of fixed communication paths will result in the need for routing algorithms which may introduce a certain degree of inefficiency. This begs the question as to the structure of the optimum topology, that is for a cross-section of applications, which topology offers the fewest problems?

1.2 System architectures

The von Neumann system architecture forms the basis for the majority of current microcomputer systems. These systems can be grouped as follows:

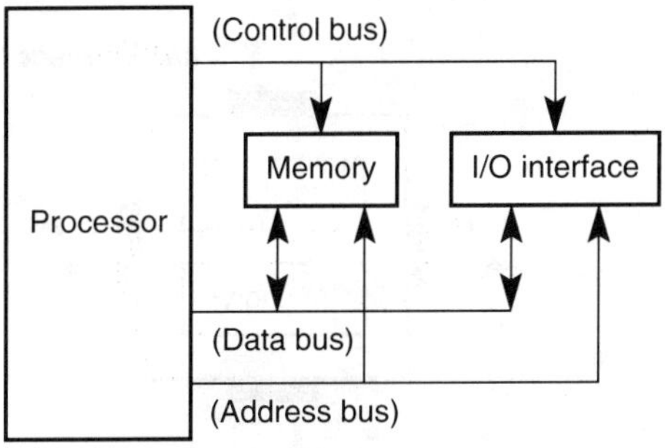

Figure 1.1 Conventional microcomputer organization.

(1) *Conventional systems* Conventional microcomputer systems have an organization as shown in Figure 1.1. The system consists of a single processor, memory and an I/O interface unit. These are implemented on separate LSI (Large Scale Integration) devices and interconnected via buses, each dealing with a particular aspect of processing.

The implementation of parallel processes on a conventional uniprocessor system requires the use of a software kernel to effect processor scheduling, interprocess communication and synchronization activities. Concurrent evaluation of the parallel processes is accomplished by executing the 'ready' processes automatically on a time slice basis, or by explicitly switching execution between the processes under program control, thus giving the parallel processing effect. As a consequence, this form of concurrent execution is generally referred to as **quasi-parallel processing**.

(2) *Transputer systems* The Inmos transputer is a complete, conventional microcomputer on a single LSI device. The design of the transputer is to facilitate the development of parallel systems based directly on CSP concurrency concepts, which are embodied in the occam language. Figure 1.2 illustrates the basic characteristics of the transputer device. To support interprocess and interprocessor communication, each transputer has four independent I/O interface links which enable simultaneous data transfers as well as point-to-point connection to other transputers. This allows multi-transputer-based systems of varying topologies and sizes to be easily configured. Each transputer is a self-contained device that uses its own resources. Therefore, the processing power, memory and communications bandwidth of such systems increases in proportion to the number of transputers incorporated within it. Figure 1.3 shows the structure of some typical multi-transputer systems.

A multi-transputer-based system allows for genuine parallel processing if each process is implemented on a separate transputer. Communication between

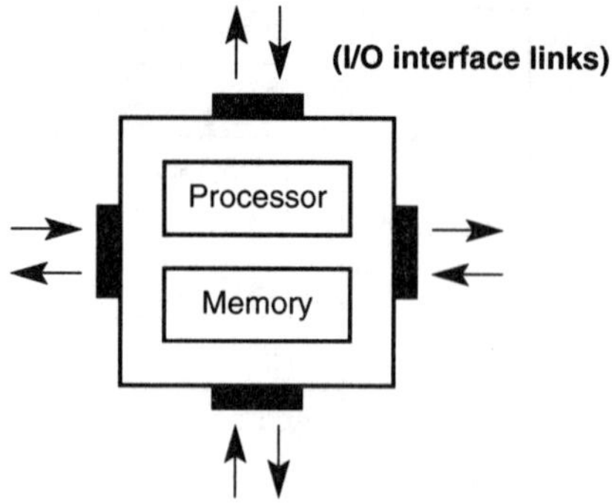

	Model		
Features	T212	T414	T800
Inter-register and bus width	16-bit	32-bit	32-bit
On-chip RAM size	2 Kbyte	2 Kbyte	4 Kbyte
Directly addressable memory space	64 Kbyte	4 Gbyte	4 Gbyte
Execution speed	10 MIPS	10 MIPS	10 MIPS

Figure 1.2 Transputer characteristics.

processes on different transputers is effected by sending unbuffered databytes via the interface links. This is accomplished when a process in one transputer outputs to a link, and a process in another has requested input from the same link; both processes wait until the transfer condition is satisfied.

Individual transputers contain a hardware kernel to enable resident occam parallel processes to be executed on a quasi-parallel basis. In situations where processes initiate communication (on the same or different transputers) or get delayed until the transputer timer reaches a certain value, the processes involved are suspended, allowing other processes waiting in the transputer to execute. After the communication or time delay is complete the suspended processes are restarted. Waiting processes are selected based on their implied priority.

Transputers therefore allow genuine and quasi-parallel processing to be implemented for any given configuration using a highly flexible and varied software structure. A major advantage is that occam programs can be developed on a single transputer and then readily targeted to a multi-transputer system of any topology.

The use of conventional microcomputers for embedded system control purposes is widespread, and the techniques employed well established. Transputers permit the highly efficient application of parallel processing for time critical control systems. In this book, the application of parallel processing

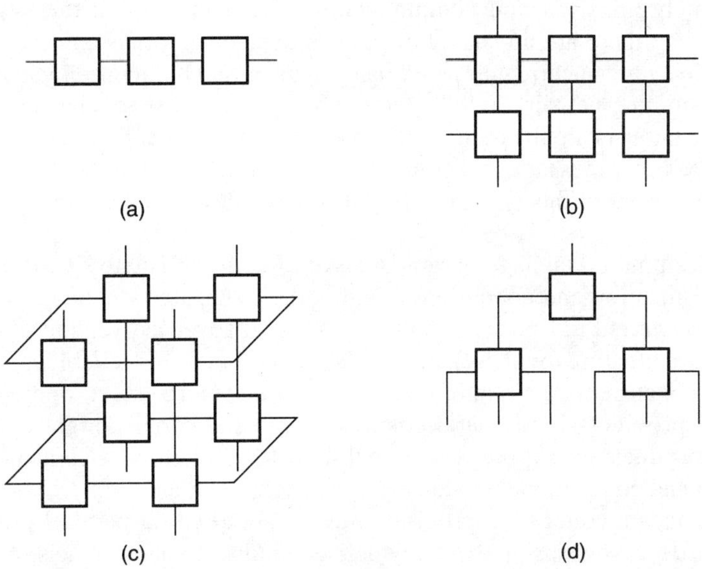

Figure 1.3 Multi-transputer system structures. (a) Linear, (b) matrix, (c) cube, (d) tree.

in embedded systems involving conventional uniprocessor microcomputer and multi-transputer-based systems will be studied.

1.3 Software developments

During the 1970s the development of concurrent programming languages came to the fore not least in studying the design of centralized and distributed operating systems, which inherently display parallel activity.

The concept of a process as a sequence of instructions which can be performed in parallel with other program instructions was generally accepted. The classification of concurrent programming languages was as follows:

(1) *Languages based on shared variables* These languages consist of active objects (processes) and passive objects (modules and monitors), with process interaction based on shared variables. If more than one process wishes to use a resource then each must be protected from interference from the other. This is known as **mutual exclusion**, and is provided for in languages of this classification. Process synchronization is also provided, as each process must be aware of the other, particularly when one process requires the result produced by another. Languages of this classification include Concurrent Pascal, Pascal Plus, Modula-2, Edison.

(2) *Languages based on message passing* These languages provide *send* and *receive* or (input/output) primitives as the primary means for process

> synchronization and communication. Each process in the system is active, there are no shared or passive objects, and thus an object is not subject to concurrent access. Such languages are highly desirable for programming network architectures. Thus processes share channels where the latter is an abstraction of a physical communication network. Two types of message passing exist: synchronous and asynchronous. Languages of this type include CSP, Gypsy, PLITS and occam.

The programming language occam is based directly on Hoare's CSP notation of communication and concurrency, and is the transputer's natural language. Occam 2 (referred to as occam in this book) is an advanced version of occam, and contains features which facilitate the development of reliable as well as efficient concurrent programs for any target transputer system topology. The language provides simple parallelism concepts and concise high-level constructs, but does not support traditional high-level features such as pointers, recursion and conventional modular program construction.

In occam concurrent programming is modelled on parallel processes which pass messages using explicitly defined unidirectional channels. All communication is synchronized, that is if a process outputs (or inputs) using a channel it will suspend execution until a process inputs (or outputs) on the same channel. As a consequence, process synchronization is accomplished, and processes executing asynchronously are still able to cooperate in an orderly fashion. Real-time programming is supported by constructs that enable the transputer timer to be read, and process execution to be delayed for specific time periods. The fundamental unit of the language is the process. Three primitive processes form the basis of occam programs, namely assignment, input and output. These may be combined to form structured processes that have sequential, parallel and alternative execution features.

The following chapters in this book look in further detail at the transputer's hardware and software, and in particular at the occam programming language, before embarking on a study of various case studies.

Chapter 2
Occam: An Introductory Course

K. Adamson and R.J. Millar

2.1 Introduction

This chapter will provide an elementary introduction to concurrent programming using the occam language. The language is regarded by Inmos as an assembly-level language providing a good level of efficiency and performance while at the same time offering the reliability of a high-level language.

The founding principle of occam is a minimalist approach which avoids any unnecessary duplication of language mechanisms. As a result, occam is a 'small' language in terms of the number of keywords and symbols. That is not to say that occam is any less capable than larger languages; it has the facility of a well-tried sequential language such as C or Fortran plus mechanisms for aspects of modern computer architecture that were impossible for the designers of early sequential languages to consider. Concurrency, alternation, communication and timers are an integral part of the occam model (Zenith, 1987; May *et al.*, 1988); an application is described as a collection of processes which operate concurrently and communicate via channels.

The popular systems language C, in common with other similar languages, was designed to 'reflect the capabilities of current computers' (Kernighan and Richie, 1978). The design and development of occam has taken a slightly different approach to this, and holds a unique position in the history of language design. The design of the Inmos transputer has been closely tied to the development of occam and the architecture involved. The transputer could be referred to as an 'occam machine', as at each stage of the transputer's development consideration has been given to the implementation of the occam model (Inmos Ltd, 1988).

Occam is also more than a programming language. It is a design formalism, just as boolean algebra is the design formalism for logic gates. A formal semantics for the language is being defined. Thus a program can be read as a set of commands for a computer to obey or, alternatively, it can be read as a predicate in an extension of the predicate calculus about which logical deductions can be made.

From the semantics are drawn a set of rules for program transformation. These rules can be used as a basis for program optimization, allowing the designer to experiment with different implementations of a program in the knowledge that the implementations are equivalent.

2.2 Architecture

Many programming languages and algorithms depend on the existence of the uniformly accessible memory provided by a conventional computer. Within the computer, memory addressing is implemented by a global communications system, such as a bus. The major disadvantage of such an approach is that speed of operation is reduced as the system size increases. The reduction in speed arises both from the increased capacitance of the bus which slows down every bus cycle, and from bus contention.

The aim of occam is to remove this difficulty; to allow arbitrarily large systems to be expressed in terms of localized processing and communication. The effective use of concurrency requires new algorithms designed to exploit this localization.

Occam is therefore based on the process model of computing, and the transputer provides a direct implementation of this model. A process is an independent computation, with its own program and data, which can communicate with other processes executing at the same time. A process can be thought of as a black box with inputs and outputs, that can communicate by message passing using explicitly defined channels (Figure 2.la). Processes can be connected together by such channels to build more complex concurrent systems. Each channel provides a one-way connection between two concurrent processes. Communication is synchronized; if a given channel is used for input in one process and output in a second process, communication takes place only when both processes are ready (Figure 2.1b). A collection of processes is itself a process, so that processes may have internal concurrency (Figure 2.1c).

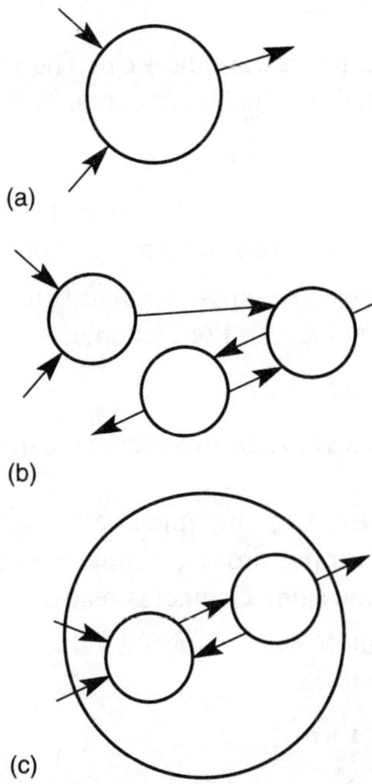

Figure 2.1 (a) A process, (b) processes connected together by channels, (c) a process with internal concurrency.

Occam enables a system to be described as a collection of concurrent processes which communicate with each other through channels. An occam program may be executed by an array of transputers, or the same program may be executed almost unchanged on a single transputer. An occam channel describes communication in the abstract and does not depend upon a particular hardware implementation, therefore a program may be written and tested without deciding where particular processes will be executed.

2.3 Fundamentals

2.3.1 Primitive processes and constructors

Occam programs are constructed from three primitive processes:

(1) *Assignment* An assignment process changes the value of a variable, as it would in most conventional languages. The special symbol used in occam is ': ='. Therefore:

```
ken := 7
```

assigns the value 7 to the variable `ken`. The value assigned to a variable could be an expression, which may contain other variables:

```
ken := 7 + mate
```

Note that '`=`' and '`:=`' are not the same. In occam '`=`' means a test for equality, whereas '`:=`' refers to an assignment.

(2) *Input* An input process inputs a value from a channel into a variable. The symbol in occam is '`?`'. For example:

```
chan1 ? server
```

This process takes a value from a channel called `chan1` and puts it into a variable `server`.

Input processes can only input values to variables, not to constants or expressions. An input process cannot proceed until a corresponding output process on the same channel is ready.

(3) *Output* An output process outputs a value to a channel. The symbol in occam is '`!`' Therefore:

```
chan2 ! ken
```

outputs the value of the variable `ken` to the channel `chan2`.

The value output to a channel can be anything that you could assign to a variable, so it may be a variable or an expression, and the expression may contain variables.

An output cannot proceed until a corresponding input process on the same channel is ready. If, during execution of a program, an input process is reached before its corresponding output process is reached, the input will wait until the output becomes ready. Should the output be reached first, it will wait for its input.

Communication is thus synchronized; at any time a process may be ready and waiting to communicate on one or more of its channels. When both an input and an output process are ready to communicate on the same channel, the value to be output is copied from the output process to the input process. These are the primitive building blocks from which occam programs are made. Each process must occupy a separate line in an occam program, and is the simplest action that occam can perform.

Occam also has two special processes, SKIP and STOP.

(1) SKIP is a process which starts, does nothing and then finishes. This is necessary in instances where the syntax of occam requires a process to be present even though nothing is required to happen. It might also be used in a partially completed program to represent a process which still has to be written.

(2) STOP is a process which starts but never proceeds or finishes. This may be thought of as representing a process which does not work. It might be used, like SKIP, to stand in for a process which has yet to be written. For example, it could be used instead of a process for handling errors, when a program is under development.

STOP should not be used to try to terminate a program. It is an instruction which can never be completed.

At this point it is important to clarify what we mean by 'finishing' when referring to processes. A process which completes all its actions is said to 'terminate'. Normally a process starts, proceeds and terminates. A process which starts but cannot proceed is said to be 'stopped'. A stopped process never terminates; for example, a process might be waiting for an input which will never happen because of a programming error. Such a process is said to be 'deadlocked'.

2.3.2 Constructors

Constructors are used to combine processes to form larger processes. The SEQuential constructor causes its components to be executed one after another, terminating when the last component terminates. The PARallel constructor causes its components to be executed concurrently, terminating only after all of the components have terminated. The ALTernative constructor chooses one component process for execution, terminating when the component terminates. Finally, IF and WHILE constructs are provided.

SEQ construction

The SEQuential construct causes the component processes to be executed one after another, terminating when the last component terminates. For example:

```
SEQ
  chan1 ? ken
  mate := ken + 1
  chan2 ! mate
```

This in effect means do in sequence, input from `chan1` to `ken`, assign `ken + 1` to `mate` and output `mate` to `chan2`. In sequence means, to be more precise, that the next process does not start until the previous one has terminated. Note that the component processes which make up the `SEQ` are all indented with respect to the word `SEQ` by two characters, which is how occam knows which processes are part of the `SEQ` construct. Note also that `SEQ` is compulsory in occam whenever two or more processes are to run in sequence, unlike more conventional languages, where the sequential execution of consecutive statements is usually taken for granted.

Replicated SEQ

Occam allows the creation of replicas of processes using a replicator index akin to an array index. For example:

```
INT var1:
SEQ index = 0 FOR 6
  channel.out ! var1 + index
```

The effect of this is equivalent to writing:

```
INT var1:
SEQ
  channel.out ! var1
  channel.out ! var1 + 1
  channel.out ! var1 + 2
  channel.out ! var1 + 3
  channel.out ! var1 + 4
  channel.out ! var1 + 5
```

that is, we create six replicas of the output process and execute them in sequence. Another new feature is the declaration of the variable `var1` to be of type `INT`, that is, integer. Types, specifications and scope will be discussed in more detail later.

The general form of a replicated SEQ is:

SEQ index = base FOR count
  process

Note that if count is zero, the replicated SEQ will act like SKIP.

PAR construction

The PARallel construct causes the component processes to start execution simultaneously, terminating only after all the components have terminated. For example:

```
PAR
  SEQ
    chan1 ? ken
    ken := ken + 1
  SEQ
    chan2 ? mate
    mate := mate + 1
```

this is interpreted as: at the same time, input from `chan1` to `ken` and then add one to the result while inputting from `chan2` to `mate` and then adding one to the result.

Notice again the indentation. The first two-character indent means that the PAR process consists of two SEQ processes. The second level of indentation shows that each SEQ is composed of two primitive processes.

The important thing to bear in mind is that in a PAR, the written order of the component processes is irrelevant as they are all performed at the same time.

We can now no longer know for sure which of the two parallel processes in the above example will finish first; it depends upon which input becomes ready first, which in turn depends upon a couple of output processes elsewhere in the program.

The beauty of occam is that this does not matter, because the PAR construction itself has a single, well-defined beginning and a single, well-defined end. We know that the two SEQ processes will start at the same time, run when their inputs become ready and then terminate.

This is a central principle of occam programming; compound processes built up from simpler processes behave just like simple processes, that is they start, perform actions and then terminate. They can in turn become the components of a still more complex process.

Replicated PAR

As for the SEQ construct, PAR may be replicated to build an array of parallel processes, any of which can be referred to by the replicator index. The general form is:

```
PAR index = base FOR count
  process
```

As an example, consider a pipeline in which data is passed from one process to another in a chain, using an array, or vector of channels, which is declared in the first line. Arrays will be dealt with in detail at a later point, however the minimal use made here should be self-explanatory:

```
[10]CHAN link:
PAR process = 0 FOR 9
  WHILE TRUE
    INT data :
    SEQ
      link[process] ? data
      link[process + 1] ! data
```

The replicator sets up nine parallel processes, each of which continually transfers values between one node in the pipeline and its neighbour. Of course, this example is not self-contained, because as it stands, the first process has nowhere to input its data from, nor has the final process anywhere to output its data to. However, we can imagine embedding this code in a larger program which supplies data to link[0] and extracts data from link[9].

2.3.3 Types and specifications

Unlike the first version of occam, occam 2 requires that objects used by a program should have a type, which must be specified before using that object in a process.

Names

Names must begin with a letter of the alphabet, may include letters, digits and the dot character, and can be of any length. Upper and lower case are distinguished by occam, so that ken and Ken are different names. Occam keywords such as SEQ, PAR, CHAN and PROC are always in upper case and are reserved.

Data types

The types available for variables in occam are:

INT	an integer,
BYTE	an integer restricted to the range 0 to 255, often used to represent characters,
BOOL	logical; either TRUE or FALSE.

In addition to these generally available types, some implementations of occam permit some or all of the following types:

INT16	16-bit integer
INT32	32-bit integer
INT64	64-bit integer
REAL32	32-bit real
REAL64	64-bit real.

Channel type and protocol

Channels are all of the type CHAN OF protocol. It is necessary to specify the data type and structure of the values that they are to carry. This is called the **channel protocol**.

A channel which carries single integer values would be specified by:

```
CHAN OF INT      chan1:
```

where the `INT` specifies the type of values which may pass along the channel `chan1`.

A protocol can be given a name in a protocol definition:

```
PROTOCOL CHAR IS BYTE:
```

which allows a channel declaration such as:

```
CHAN OF CHAR screen:
```

This simple protocol idea can be extended to a sequential protocol which specifies a communication protocol consisting of a sequence of simple protocols. For example:

```
PROTOCOL COORDS IS INT; INT:
```

Any channel declared with the `COORDS` protocol passes pairs of integer values; for example:

```
mouse ? x; y
```

Additionally, a special protocol, called a **counted array** protocol, enables communication of arrays of values where the number of array components is not known until the communication takes place. Declaration:

```
CHAN OF INT::[]BYTE string:
```

Used as:

```
string ! 15 :: "Variable length"
```

and

```
string ? len :: buffer
```

Obviously, the size of '`buffer`' must be at least the value of '`len`'.

For completeness, variant protocols are possible:

```
PROTOCOL ANALYSIS
  CASE
    word;          INT
    fourchars;     [4]BYTE
    error;         INT16::[]BYTE
    halt

:
```

declares a protocol called `ANALYSIS` which can support the communication of data in any of the named sequential formats '`word`', '`fourchars`', '`error`' and '`halt`'.

In use, the tag (that is, the name of the sequential format) must be sent first before the data format, for example

```
chan1! fourchars; "SIZE"
```

Notice also the use of only the tag to convey information:

```
chan1! halt
```

Input from such a channel is slightly more complex as illustrated by the following example:

```
chan1 ? CASE
  word; X
    ... a process associated with this input
  fourchars; array
    ... similarly
  error; len :: message
    ... handle error
  halt                  -- do nothing perhaps
```

If a tag is input which does not match one of the CASEs, the process behaves like STOP.

Timer type

The type TIMER allows the creation of timers which can be used as clocks by processes. Timers will be discussed in more detail later.

Characters and strings

Occam does not have any type CHAR or STRING to represent alphabetic characters or words. Instead, characters are represented as numbers of type BYTE and strings as arrays of numbers of type BYTE.

2.3.4 Boolean operators

A boolean value is produced as the result of a test performed by a comparison operator. The following operators are available in occam:

=	equal to
<>	not equal to
>	greater than
<	less than
>=	greater than or equal to
<=	less than or equal to.

Note that brackets are needed to remove ambiguities whenever more complicated tests are made by combining two or more comparisons.

Occam also provides the boolean constants, TRUE and FALSE, which may be used anywhere that a test could be used, and the standard boolean operators AND, OR and NOT.

Bit operators

To allow low-level operations on the individual bits in a value, occam provides the following bit operators:

`/\`	bitwise and
`\/`	bitwise or
`><`	bitwise exclusive or
`~`	bitwise not
`<<`	left shift
`>>`	right shift.

2.3.5 Constants

In occam a name may be given to a constant value by using the specification:

VAL type name IS value:

so it is possible to write,

```
VAL INT hours IS 24, minutes IS 60:
```

The type can be omitted when it is obvious from the value:

```
VAL year IS 365:
```

Possible ambiguities over BYTE and INT, or REAL32 and REAL64, are resolved by explicitly specifying the type of the value:

```
VAL empty IS 0 (BYTE):
VAL pi IS 3.14159 (REAL32):
```

The first of these is equivalent to:

```
VAL BYTE empty IS 0:
```

Note the colon, which is used to end all the different kinds of specification. The colon joins a specification to the process which follows it.

2.3.6 Scope

In occam, variables, channels and other named objects are local to the process which immediately follows their specification. What this means is that the object to which the name refers effectively does not exist inside any other process. For example:

```
PAR
  INT ken:
  SEQ
    chan1 ? ken
    ... further processes
  INT mate:
  SEQ
    chan2 ? ken
    ... further processes
```

An error will be reported, as `ken` exists only inside the first `SEQ` and `mate` exists only inside the second `SEQ`. The second `ken` will therefore look to occam like an unspecified variable.

The colon which ends a specification in effect joins the specification to the process which follows it and, to reinforce the connection, specifications are indented to the same level as the process.

Consider now the earlier `PAR` example with correct declarations:

```
CHAN OF INT chan1, chan2:
PAR
  INT ken:
  SEQ
    chan1 ? ken
    ken := ken + 1
  INT mate:
  SEQ
    chan2 ? mate
    mate := mate + 1
```

Now the channels `chan1` and `chan2` are known throughout the `PAR` process; we could legally refer to either of them in either of the SEQs. On the other hand, `ken` and `mate` are known only within their respective SEQs.

Initialization

In occam the value of a variable is unassigned until it has either input a value or has been assigned a value. Furthermore, the value of a variable has meaning only during the execution of the process for which it has been declared. Once the process has terminated the variable no longer has a well-defined value. If the process is to execute again it is important that the variable has again been assigned a value. If you need a variable to keep its value from one execution of a process to another, declare it in an outer scope, that is, before a process which contains the process which is being repeatedly executed.

2.3.7 Operators

The basic arithmetic operators available are:

a + b	add b to a
a – b	subtract b from a
a * b	multiply a by b
a / b	divide a by b
a REM b	remainder when a is divided by b
a \ b	alternative form of REM.

All operators have the same precedence and associativity in occam, so parentheses must be used in complex expressions to enclose component operations and allow them to be treated as single operands. This also establishes the order of evaluation. For example:

(2 + 1) * (3 + 4)	result is 21
3 + (3 * (2 + 6))	result is 27
4 + 6 * 8 + 3	illegal
(1 + (2 * 3)) + 6	result is 13
(1 + (2 * 3) + 6)	illegal

Modulo arithmetic operators

For integers only, a further set of operators is provided which permits modulo arithmetic. The modulo operators PLUS, MINUS and TIMES permit addition, subtraction and multiplication modulo 2^n–1, where n is the number of bits in an INT.

2.3.8 Communicating processes

Communication between parallel processes is the essence of occam programming. At its simplest it requires two processes executing in parallel and a channel joining them:

```
INT a:
CHAN OF INT chan1:
PAR
  chan1 ! 7
  chan1 ? a
```

This simple program merely outputs the value 7 from one process and inputs it into the variable a in the second.

Note that communication between the component processes of a PAR must only be done using channels. Occam does not allow the passing of values between parallel processes by the use of a shared variable.

If a component of a PAR contains an assignment or input to a variable, then the variable must not be used at all in any other component:

```
INT a, b:
PAR
  SEQ
    a := 7
    ... further processes
  SEQ
    b := a        -- ILLEGAL
    ... further processes
```

Keeping variables local to component processes and using channels to communicate values is the right way to do it.

This seemingly severe restriction proves to be a major source of error when first programming in occam, but is necessary for the following reason. Parallel processes run at the same time, and in general they run asynchronously, that is, at their own pace, only coming into synchronization with each other when forced to by communication over a channel.

If the language allowed one parallel process to read from a variable which has its value altered in another parallel process, what value will be read? It depends upon whether or not the other process has altered it yet, and this cannot be known, since the processes are asynchronous. Again, consider the problems if the altering process chooses to alter the variable's value at the precise moment that the second process is reading it!

Therefore, in occam variables are used for storing values, while channels are used for communicating values. Thus an occam channel provides a one-way link between a pair of processes; in a communicating PAR construct this means that:

(1) Only two component processes of the PAR may use any particular channel, one as the sender and the other as the receiver:

```
CHAN OF INT chan1:
PAR
  SEQ
    chan1 ! 7
  INT a:
  SEQ
    chan1 ? a
  INT b:    -- ILLEGAL! two processes
  SEQ       -- inputting from the same channel
    chan1 ? b
```

(2) The sender process must only contain outputs to the channel and the receiver must only contain inputs from the channel:

```
CHAN OF INT chan1:
PAR
  SEQ
    chan1 ! 7
  INT a:
  SEQ
    chan1 ? a      -- ILLEGAL! input and
    chan1 ! a + 1  -- output from the same
                   -- channel in the same
                   -- process
```

For two-way communication between two processes we would need two channels:

```
CHAN OF INT chan1, chan2:
PAR
  INT a:
  SEQ
    chan1 ! 7
    chan2 ? a
  INT b:
  SEQ
    chan1 ? b
    chan2 ! 40
```

The effect is that each process sends a value to the other; `b` will have the value 7 and `a` will have the value 40. The order of the inputs and outputs in each `SEQ` is crucial here. For example:

```
CHAN OF INT chan1, chan2:
PAR
  INT a:
  SEQ
    chan2 ? a
    chan1 ! 7
  INT b:
  SEQ
    chan1 ? b
    chan2 ! 40
```

This program would never terminate – DEADLOCK. Both SEQs are waiting for an input; however, since each is waiting for the other to output, neither can proceed. Swapping the input and output in either process resolves the deadlock.

2.3.9 Conditional processes

Occam provides a form of conditional choice by means of the construct IF, which takes any number of processes, each preceded by a test, and builds from them a single process. The conditions are tested sequentially, and the first one which is TRUE is executed. For example:

```
IF
  temp = 1
    chan1 ! a
  temp = 2
    chan2 ! a
  temp = 3
    chan3 ! a
```

The effect of this piece of code is to output the value of the variable `a` on `chan1`, `chan2` or `chan3`, depending upon whether the value of `temp` is l, 2 or 3. Note that if the value of `temp` is anything other than l, 2 or 3 the effect of this `IF` process is equivalent to `STOP`. The program can only proceed if one of the choices is executed.

It would generally be unacceptable to have the program stop if `temp` is neither l, 2 or 3. How could this difficulty be avoided?

```
IF
  temp = 1
    chan1 ! a
  temp = 2
    chan2 ! a
  temp = 3
    chan3 ! a
  TRUE
    chan4 ! a
```

Now `a` will be output on `chan4` if `temp` has any value other than l, 2 or 3 because the test on the last choice is always true.

If we wanted nothing to happen at all when `temp` was neither l, nor 2 nor 3 we could use the following:

```
IF
  temp = 1
    chan1 ! a
  temp = 2
    chan2 ! a
  temp = 3
    chan3 ! a
  TRUE
    SKIP
```

This provides one example of the use of `SKIP`. Occam requires some sort of process after the condition.

To make more complex choices, IFs can be nested by using an IF process in a choice of another IF construct:

```
IF
  temp = 1
    chan1 ! a
  temp = 2
    IF
      a = 7
        chan2 ! a
      TRUE
        chan3 ! a
  TRUE
    SKIP
```

If the value of `temp` is 2 the effect of this code fragment is to output the value of `a` on `chan3` if `a` has any value other than 7.

Alternative processes

Whereas the IF construct enables us to choose different processes according to the values of variables in the program, the ALTernative construct allows us to make choices which depend upon the state of channels. The component parts of ALT, referred to as **alternatives**, are combined by ALT into a single construct; however, they are slightly more complicated.

Therefore, each component process has a **guard**, which is an input with an optional condition. The earliest process which is ready to be executed is chosen, then the guard is executed followed by the guarded process. If several alternative guards are ready an arbitrary one is chosen. For example:

```
WHILE going
  ALT
    buffer.in ? ch
      buffer.out ! ch
    stop ? ANY
      going := FALSE
```

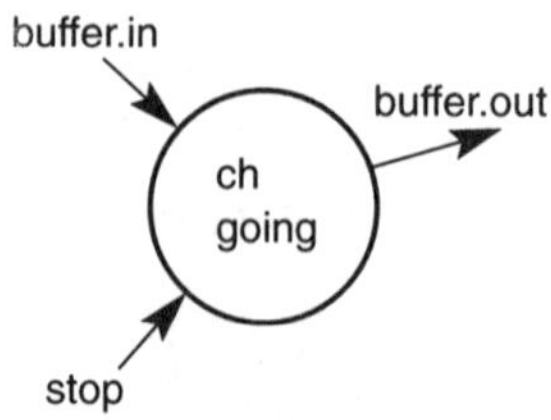

The variable `going` is a boolean variable which is initially set to TRUE. The process is terminated by a message input on the `stop` channel. `buffer.in` and `stop` are therefore channels feeding the process, `buffer.out` is the process output channel. The process behaves like a simple buffer, reading in on `buffer.in` and outputting on `buffer.out` until a STOP message is received. The guards in this case are effectively the availability of an input on the respective channels. A process which is guarded by an input is not executed unless the process at the other end of the channel is waiting to output.

As mentioned above, an ALTernative may start with a test in addition to an input. If this is the case, the associated process can only be chosen if its input is the first to be ready and the test is TRUE. Occam makes this easy to remember by using the '&' sign, as in:

```
CHAN OF INT chan1, chan2, chan3:
INT a:
ALT
  (temp < 0) & chan1 ? a
    ... process 1
  (temp = 0) & chan2 ? a
    ... process 2
  (temp > 0) & chan3 ? a
    ... process 3
```

As with IF, `ALT` behaves like STOP if no alternatives can ever be executed. Also like IF, an ALT can be nested inside an outer ALT.

The ALT is an extremely powerful construction, allowing complex networks to be merged and switched in a straightforward way.

2.3.10 Repetitive processes

We may wish to have a process executed repeatedly, until some condition is satisfied. One way in which to achieve this is to use the construction WHILE, which includes a test such as a < 0 or temp = 70. The resulting process is executed until the boolean expression evaluates to FALSE. For example:

```
INT a:
SEQ
  a := 0
  WHILE a >= 0
    SEQ
      buffer.in ? a
      buffer.out ! a
```

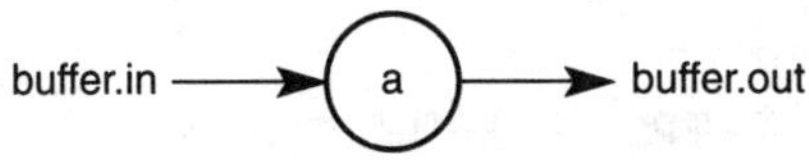

This will continue to read values from the channel `buffer.in` and send them to `buffer.out` so long as the value is not less than zero. Every time the inner `SEQ` process terminates, the `WHILE` process will be performed again and the test repeated. This continues so long as the test result is TRUE, that is so long as `a` is greater than or equal to zero.

2.3.11 Example

The following example is intended to illustrate the use of these more difficult constructs. Suppose that we want to write a program to control a heating system equipped with digital controls. To control the heat output we have two digital controls, hot and cold. Pressing hot increases heat output by one unit, and likewise pressing cold reduces it.

We have two occam channels, also called `hot` and `cold`, which produce an input whenever a button is pressed. A third channel, `regulator`, has the function of transmitting a value to the regulator mechanism on the heat source.

The processes which increase or decrease the temperature are easily written:

```
SEQ
  temp := temp + 1
  regulator ! temp
```

and

```
SEQ
  temp := temp - 1
  regulator ! temp
```

Now the program needs to decide which button was pressed most recently, and hence which action to take. Combining the two processes in an `ALT` will achieve this. The `WHILE` construct is used to continually scan the buttons:

```
INT temp, any:
SEQ
  temp := 0
  regulator ! temp
  WHILE TRUE
    ALT
      hot ? any
        SEQ
          temp := temp + 1
          regulator ! temp
      cold ? any
        SEQ
          temp := temp - 1
          regulator ! temp
```

The above program would work, but the control panel buttons are repeatedly interrogated and the process never terminates. To ensure satisfactory termination an extra channel can be added to communicate the status of the control panel OFF button, and the boolean constant `TRUE` can be replaced by a boolean variable:

```
BOOL operational:
INT temp, any:
SEQ
  operational := TRUE
  temp := 0
  regulator ! temp
  WHILE operational
    ALT
      hot ? any
        SEQ
          temp := temp + 1
          regulator ! temp
      cold ? any
        SEQ
          temp := temp - 1
          regulator ! temp
      off ? any
        operational := FALSE
```

As a further refinement we can add tests to the temperature increase and decrease processes to limit the values to the range which the regulator can accept (consider a possible range of from 0 to 85).

We could merely add the tests `temp < 85` and `temp > 0`. As a matter of good programming style, it would be better to define the limits as named constants at the beginning of the program; therefore, if a new heater is installed, revision of the temperature limits is straightforward:

```
VAL min.temp IS 0, max.temp IS 85:
BOOL operational:
INT temp, any:
SEQ
  operational := true
  temp := min.temp
  regulator ! temp
  WHILE operational
    ALT
      (temp < max.temp) & hot ? any
        SEQ
          temp := temp + 1
          regulator ! temp
      (temp > min.temp) & cold ? any
        SEQ
          temp := temp - 1
          regulator ! temp
      off ? any
        operational := FALSE
```

Note that the channels `regulator`, `hot`, `cold` and `off` have not been declared in the program. That is because they connect to the hardware rather than to other occam processes; ultimately, they represent physical bits of wire connected to a button.

In a real control program there might be many other things for the program to do besides reading temperature buttons; for example, there may be readings coming from room sensors, or radiator sensors which will affect the regulator. Processes to carry out these tasks could be combined with the above program using a `PAR` in place of the main `SEQ` so that they all proceed at the same time.

2.3.12 Abbreviations

One of the most powerful and useful features of occam is the abbreviation. Abbreviations may be used to give a name to an expression in occam. For example:

```
VAL seconds IS 60*((60*hours) + mins):
```

which defines `seconds` to be shorthand for the value of the expression on the right.

The fact that this specification ends with a colon signifies that abbreviations are local, like other occam objects, and their scope is the following process.

If an abbreviation, such as in the above example, contains variables on its right hand side, then these variables should remain constant throughout its scope. The full form of an expression abbreviation contains a type specifier before the name:

VAL INT name IS expression:

but this can be omitted, leaving occam to deduce the type from the type of the right hand side.

2.3.13 Procedures

A procedure is a process with a name. The keyword PROC and a name is followed by a process which is called the **procedure body**.

The body of a procedure is executed whenever its name is found in a program; such an occurrence of the name is called an **instance** of the procedure.

A procedure definition looks like this:

```
PROC delay ( )                  -- procedure heading
  VAL interval IS 100:          --
  INT n:                        --
  SEQ                           -- procedure body
    n := interval               --
    WHILE n > 0                 --
      n := n - 1                --
:
INT y:                          -- main process
SEQ                             -- starts here
  input1 ? y
  delay ( )                     -- instance
  output1 ! y
  delay ( )                     -- instance
  input2 ? y
  delay ( )                     -- instance
  output2 ! y
```

All this procedure does is count downwards from 100 to 0, so it could be used as a crude way of introducing a time delay into a program (the proper way is to use timers).

Procedures obey the same scope rules as other occam objects such as names and variables; the procedure is only known throughout the process which immediately follows it, to which it is linked by the final '**:**'.

Parameters

In the delay example above, the length of the delay is fixed at 100 in the text of its definition, and it can only be changed by editing the program. A more flexible way would be to pass the delay length as a parameter:

```
PROC delay (VAL INT interval)
  INT n:
  SEQ
    n := interval
    WHILE n > 0
      n := n - 1
:
INT y:                              -- main process
SEQ                                 -- starts here
  input1 ? y
  delay (100)
  output1 ! y
  delay (2000)
  input2 ? y
  delay (700)
  output2 ! y
```

The name interval in the definition of delay is called a **formal parameter**. Formal parameters may be of any type, including CHAN, and a type specification is compulsory.

A procedure can have any number of formal parameters, which must be separated by commas in the definition heading. The actual parameters of an instance are similarly separated by commas, and must correspond in number, position and type with the formal parameters of the PROC.

Parameter passing

When a variable is passed as an actual parameter to a procedure in occam, that variable effectively replaces the formal parameter throughout the procedure. Anything which the procedure does to the formal parameter is also done to the variable, which may in consequence have its value changed. This is in contrast to the call-by-value convention commonly used in other programming languages in which the actual parameter is used as the initial value of the formal parameter, which then behaves as a local variable of the procedure. The occam convention is more equivalent to Pascal's call-by-reference (or VAR) parameters.

We sometimes require that a procedure should not alter the value of a variable which has been passed to it as a parameter. One way in which to achieve this is to assign the value of the parameter to a local variable within the PROC, and to perform any operations on this local copy. If we need only the original value of a variable in the body of a procedure (that is, the formal parameter is never altered by assignment or by input) then we can say explicitly that only the value is to be passed, using VAL. This was used in the above delay procedure.

We can rewrite the earlier heating control program to exploit the properties of procedures:

```
VAL more IS 1, less IS -1:
BOOL operational:
INT temp,any:
PROC change.temp (VAL INT step)
  SEQ
    temp := temp + step
    regulator ! temp
:
SEQ
  temp := 0
  regulator ! temp
  operational := TRUE
  WHILE operational
    ALT
      hot ? any
        change.temp (more)
      cold ? any
        change.temp (less)
      off ? any
        operational := FALSE
```

Note that the body of the `PROC` uses the variable `temp` even though it is not declared anywhere within the procedure, either as a local variable or a formal parameter; `temp` is referred to as a **free variable** with respect to the `PROC` `change.temp`. Most importantly, it retains its value from one call of the procedure to the next, which is why it is used here. Note that `temp` has to be declared somewhere before the `PROC` definition, otherwise it would be rejected as an undeclared name. In general, a variable is free with respect to a procedure when the procedure is defined within the scope of that variable.

2.3.14 Functions

A **function** gives a name to a special kind of process which returns a result, called a ‘value process’. Occam functions are side-effect free; when used they will have no effect upon any other part of the program. The definition of a function has the following general form:

type FUNCTION name ({,formal parameters})
  specification:
  VALOF
    Process
    RESULT expression
:

For example, consider the following function which returns the value of the larger of two integers:

```
INT FUNCTION max (VAL INT x,y)
  INT answer:
  VALOF
    SEQ
      IF
        x > y
          answer := x
        y > x
          answer := y
        x = y
          answer := x
  RESULT answer
:
```

The rules for constructing functions and value processes are very strict:

- the formal parameters of a function can only be VAL parameters,
- PARallel and ALTernation constructs cannot be used within the function,
- input and output must not be used within the function,
- you can only assign to variables declared within the scope of the function; 'free variables' can be read but not assigned to.

A function returns a value and is defined as an operand, so functions can appear wherever an expression would appear. For example:

```
z := max (x,y) * 4
```

2.4 Arrays

Unlike earlier versions of occam, occam 2 supports multidimensional arrays. Generally, an array is a set of elements of the same type. An array may have one or more subscripts – the number of subscripts is referred to as the **dimension** of the array. An element of an array is specified by giving the value of each of the array subscripts, and is usually known as a **component** of the array.

2.4.1 Array types

Array variables in occam are declared in the same way as single variables of any type, but with the number of components in each dimension prefixing the type declaration. For example:

[7]INT ken:	–	an array of seven integers called ken
[40][40]BYTE twod:	–	a 2-dimensional byte array
[10][10][10]INT threed:	–	a 3-dimensional integer array
[1000]REAL32 results:	–	a vector of 1000 reals
[7]CHAN switch:	–	a vector of 7 channels.

An array can be referenced by name to transmit it to another process:

```
PAR
  chan1 ! twod
    processes
  chan1 ? matrix
```

This will send the entire array `twod` to another process, provided that `matrix` has also been declared to be of type `[40][40]BYTE`.

Note that the sizes of arrays in occam must be fixed at compile time and cannot be assigned or altered during execution.

The convention for referring to particular components of arrays in occam is to specify the array name followed by a suffix or suffixes giving the particular values of the array subscripts in brackets.

For example, consider a process to output a one-dimensional array to a channel:

```
CHAN output:
[40]INT data:
SEQ
  element := 0
  WHILE element < 40
    SEQ
      output ! data[element]
      element := element + 1
```

This could also be achieved using a replicator:

```
CHAN output:
[40]INT data:
SEQ element = 0 FOR 40
  output ! data[element]
```

Arrays may be used as parameters to procedures. In occam, the size of an array parameter need not be declared when the PROC is specified, which allows the passing of an array of any size as an actual parameter, provided that it is of the correct data type. The type specifier for an array formal parameter looks like an array type with the size omitted. However, since a procedure may need to know the size of such an array, there is a special operator, SIZE, in occam which returns an INT value which is the size of the array. For example:

```
PROC input (CHAN in, []INT store)
  INT element:
  SEQ
    element := 0
    WHILE element < (SIZE store)
      SEQ
        in ? store[element]
        element := element + 1
:
```

2.4.2 Array segments

An array segment consists of a set of consecutively numbered components of an array, and is of the form:

[array FROM subscript FOR count]

A segment can itself be treated as an array. Thus defining the array `[40]INT data` we may treat the segment

```
[data FROM 7 FOR 10]
```

as an array of 10 components, starting with `data[7]` and ending with `data[16]`.

Array segments can be input, output or assigned to in the same manner as arrays, provided that the expression which is assigned is always an array of the same type and size as the segment. For example:

```
[data FROM 7 FOR 10] := [data FROM 0 FOR 10]
[data FROM 7 FOR 10] := store
```

where `store` has been declared as `[10]INT`.

2.4.3 Abbreviations

We have already discussed the use of abbreviations for constants and constant expressions. The description of array segments can also be much simplified by using abbreviations. For example:

```
dat IS [data FROM 7 FOR 10]:
```

enables us to consider `dat` to be an array of size 10, with subscripts running from 0 to 9, where, for example, `dat[2]` is identified with `data[9]`, and so on.

Notice that VAL is not used, as it is when abbreviating a constant or expression; we are not merely naming a value here. An abbreviation for an array, array component, array segment or simple variable behaves just like the object itself.

In other words, it is permissible to change the value of the object referred to by the abbreviation name, either by assignment or input. We could write the previous example more succinctly as:

```
dat := store
```

What is not permissible, in the case of an abbreviation for an array component or array segment, is to alter which component or components are referred to. For example, if we declare an abbreviation:

```
element IS data[i]:
```

then it is illegal to change the value of the. subscript `i` within the scope of `element`; `element` must always refer to the same component of data, defined by the value of `i` at the time it was specified.

Abbreviations may also reward the programmer with performance benefits, as well as improving the conciseness of code. If, instead of subscripting an array, we use an abbreviation for an array component within a loop:

- the compiler recognizes that the subscripts are constant and so does not compile run-time checks,
- the address of the array component becomes local to the loop process rather than global, and occam processes handle local data faster.

An abbreviation may also be used to set up an array constant:

```
VAL months IS [1,2,3,4,5,6,7,8,9,10,11,12]:
```

The components of `months` can be accessed in the usual way, so `months[2]` is 3.

2.4.4 Tables

Occam provides a powerful means of generating an array value, called a table. If `Z` is a variable of type `INT`, and its current value is 7, the table:

```
[Z, Z+1, Z+2, Z+3, Z+4]
```

generates an array of type `[5]INT` and value:

```
[7,8,9,10,11].
```

This array can be assigned to a variable:

```
INT Z:
[5]INT speed:
SEQ
  ...
  speed := [Z, Z+1, Z+2, Z+3, Z+4]
```

or it may be referenced by subscript in the usual way:

```
chan1 ! [Z, Z+1, Z+2, Z+3, Z+4][3]
```

or it may be abbreviated to a name for later use:

```
INT Z:
...
VAL speed IS [Z, Z+1, Z+2, Z+3, Z+4]:
SEQ
  ...
  chan1 ! speed[3]
```

2.5 Replicators

Earlier we introduced the use of replicators with the SEQ and PAR constructs. Now, armed with more details of the properties of arrays in occam, two further replicated constructs, IF and ALT, are discussed.

2.5.1 Replicated IF

The general form is:

```
IF index = base FOR count
  choice
```

where choice consists of a condition followed by a process. For example:

```
IF element = 0 FOR 10
  data[element] = 1
    data[element] := 0
```

whose effect is simply to test the first 10 elements of the `data` array and to replace the first one found to be 1 by 0. Note, however, that if no 1 is found the process will be stopped. The replicated construct does not admit (sensibly) a concluding TRUE SKIP. Most uses of replicated IF therefore involve nesting it within an outer IF. In this case, the behaviour if no conditions are met is more appropriate, the next choice in the outer IF is considered:

```
IF
  IF element = 0 FOR 10
    data[element] = 1
      data[element] := 0
  TRUE
    SKIP
```

This will now SKIP and continue the program if no 1s are found in the array. This construct could be used for searching for the first occurrence of a given character in a string, as illustrated by the function:

```
INT FUNCTION getchar (VAL BYTE char,
                                          VAL []BYTE string)
  VAL INT nomatch IS -1:
  INT result1:
  VALOF
    IF
      IF position = 0 FOR SIZE string
        string[position] = char
          result1 := position
      TRUE
        result1 := nomatch
      RESULT result1
:
```

The function returns –1 to indicate the character has not been found.

2.5.2 Replicated ALT

A **replicated ALT** consists of a number of identically structured alternatives, each of which is triggered by input from a channel; it has the general form:

ALT index = base FOR count
    alternative

where alternative is a guard followed by a process. The effect is to monitor an array of channels, and it may therefore be used to construct a multiplexer:

```
PROC multiplexer ([]CHAN in, CHAN out, disable)
  INT any, data:
  BOOL operational:
  SEQ
    operational := TRUE
    WHILE operational
      ALT
        ALT x = 0 FOR (SIZE in)
          in[x] ? data
            output ! data
        disable ? any
          operational := FALSE
:
```

Replicated ALT provides an elegant and concise way to build and control switching networks of all shapes and sizes. The array of channels could carry readings from a battery of instruments, or control information from sensors distributed throughout a machine.

2.6 Real-time programming

2.6.1 Timers

Timing in occam is provided by declaring named objects of the type TIMER. A timer behaves like a channel which can only provide input representing the current time as a value of type INT. For example:

```
TIMER clock:
  INT time:
    clock ? time
```

Note that the 'ticks' of this clock will vary from one implementation of occam to another, depending upon the hardware on which it is running. On the Inmos transputer the ticks will be in units of (input clock rate)/(5 $*$ 64) which will normally work out at 64 microseconds per tick; it is, however, essential to consult the hardware manuals for a given system. Whenever the value of time exceeds the maximum value that can be represented by an INT it will become negative and begin to count backwards towards zero (in 2's complement signed arithmetic). With a 64 microsecond tick and a 16-bit INT this would happen approximately every 4.2 seconds. Time differences must therefore exclusively be calculated using the modulo arithmetic operators, and a long interval can only be timed by breaking it into a series of shorter intervals.

2.6.2 Delays

A delayed input is an input from a timer which cannot proceed until the time reaches a stated value. The operator AFTER followed by an expression representing a time is used to cause the delay. For example:

```
PROC delay (VAL INT period)
  TIMER clock:
  INT present.time:
  SEQ
    clock ? present.time
    clock ? AFTER present.time PLUS period
:
```

An instance of this procedure, say `delay (7000)`, would pause for 7000 ticks before terminating. Note that no variable has its value changed, the value from `clock` is only compared with the value of the expression. Note also that several components of a PAR are allowed to input from the same timer, which would be strictly forbidden for an ordinary channel.

A delayed input could be used in an ALT to provide a real-time wait:

```
TIMER clock:
VAL time.limit IS 700:
INT present.time:
SEQ
  clock ? present.time
  INT Z :
  ALT
    input ? Z
      ... further
      ... processes
    clock ? AFTER present.time PLUS time.limit
      warning ! (17 :: "Timeout on input!")
```

This process will send the timeout warning message if channel input does not produce an input within the prescribed time of 700 ticks.

2.6.3 Priority

It is sometimes useful, particularly in real-time programming, to be able to assign priorities to processes in a well-defined way. We saw earlier in the case of the ALT construct that when the inputs guarding two alternative processes become ready simultaneously, occam makes an arbitrary choice between the two. The problem is that the programmer cannot predict what will happen in such a case. Occam allows the assignment of priorities for the ALT and the PAR constructs by preceding the keywords by PRI. In either case, the component processes are assigned a priority which corresponds to the textual order in which they appear in the program.

PRI ALT

As already indicated, when two processes become ready simultaneously, the process with the higher priority will be executed. An example of the use of this construct is the routine below which guarantees that a channel carrying an important signal is examined:

```
WHILE operational
  INT any:
  PRI ALT
    reset ? any
      operational := FALSE
  TRUE & SKIP
    ... main body
```

Here the `PRI ALT` forces the program to check the channel `reset` because of its higher priority. The `TRUE & SKIP` option is always ready, and if used in an ordinary ALT this path could be taken at every cycle without reset ever being examined.

PRI PAR

Assigning priorities to the component processes means that processes with a lower priority can only proceed if no higher priority process is able to proceed. For example:

```
PRI PAR
  SEQ
    res1.in ? data1
    res1.out ! data1
  SEQ
    res2.in ? data2
    res2.out ! data2
```

Here the second `SEQ` cannot proceed, even when ready, unless the first is waiting for its input or output.

As an example, it might be the case that it is more important to relay any messages to an external device than to continue computation in a program. PRI PAR can be used to ensure that computation only proceeds when there is no message waiting to be input or output:

```
PRI PAR
  WHILE TRUE
    [string.length]BYTE message:
    SEQ
      input ? message
      SEQ i = 0 FOR string.length
        output ! message[i]
      ... main body
```

This program has assumed (for illustrative purposes) that the external device on the other end of the channel output requires serving with data one byte at a time.

If PRI PAR is used, the question of buffering may well arise. It is pointless to run a process at high priority to service an impatient device if, while servicing that device, it can be kept waiting to communicate with another process. A high priority process of this kind should have all such communications with other processes buffered so that data can be sent without delay. The size of the buffer needed would be tuned to the actual timings of the processes involved:

```
PRI PAR
  CHAN buffer:
  PAR
    WHILE TRUE
      [string.length]BYTE message:
      SEQ
        input ? message
        buffer ! message
    WHILE TRUE
      [string.length]BYTE message:
      SEQ
        buffer ? message
        SEQ i = 0 FOR string.length
          output ! message[i]
          ... main body
```

In this revised example, the input from the channel input has a buffer added. This reduces the chances of the service routine having to wait to input the next message, which would cause it to delay service to the device.

As a general rule, PRI PAR should only be used when it is essential to impose explicit priority. Since priority does not impose upon the logical structure of a program, it is really a configuration issue rather than an occam programming issue, and thus should be left until last, when the overall program logic has been established and the program works.

2.7 Characters and strings

In occam single characters may be written as the character contained in single quotes, for instance 'a'. Such a constant is translated into a one byte number by occam.

A text string may be written in double quotes, for instance "this is an exciting course". Occam translates such a string constant into an array of type [*n*]BYTE, where *n* is the length of the string. Some special codes of importance are represented as:

*c carriage return
*n newline
*t horizontal tabulate
*s space
*' quotation mark
*" double quotation mark
** asterisk

More generally, any character can be represented by an asterisk followed by its code in hexadecimal notation.

String constants can be assigned to arrays and manipulated in a variety of ways. Single characters can be accepted from them by ordinary subscripting, or whole parts of the string may be selected as an array segment; for example, the function getchar discussed earlier under REPLICATORS which returns a pointer to the first occurrence of a given character found in a string.

An occurrence of this function in the assignment:

```
pos := getchar ('e', "creative")
```

causes `pos` to become 2, which is the subscript of the letter `e` in the string `"creative"`.

As a further example, the procedure below outputs a string in upper case, converting any lower case characters to upper case before output. Lower case ASCII characters have codes which are larger by 32 than the upper case character code (for example, 'a' is ASCII code 97, while 'A' is 65). Expressed in binary, this means that the sixth bit is 1 in the lower case version but is 0 in the upper case. Therefore, to convert to upper case we need to bitwise AND (/ \) with binary 11011111, that is hexadecimal DF:

```
PROC uppercase.output (CHAN OF BYTE output,
                                    [] BYTE string)
  INT x:
  SEQ
    x := 0
    WHILE x < (SIZE string)
      SEQ
        output ! string[x] /\  DF
        x := x+1
:
```

In this procedure the string is output as a stream of BYTES and no size information is sent. The inherent problem here is that any receiving process would not know when the stream had ended, and could attempt to keep inputting characters once the `WHILE` in the output procedure had terminated. This would result in deadlock. A simple solution would be either to send the size of the string as the first item, or alternatively send a special ‘end-of-stream’ character.

2.8 A pipelined sort program

The principle of this program is that a stream of unsorted numbers is passed into a pipeline which has as many parallel processes as there are numbers. Each replicated process has the local variables **largest** and **nextnumber**.

As a number enters a new process, it will be compared with the value in largest. If it is not greater than largest it will be passed straight on to the next process in the pipe. If it is greater than largest it will replace the value of largest, and the previous value will be sent on to the next process. When all the numbers have been through the process, the final value of largest is passed on.

In the first process the largest number migrates to the last position in the stream; in the second process the second largest migrates to the second from last, and so on. For example:

First Process

unsorted	largest	next number	passed on
52847 –>	7	4	4
	7	8	7
	8	2	2
	8	5	5
	8	finished	

Second Process

5274 –>	4	7	4
	7	2	2
	7	5	5
	7		
524 –>	and so on		

Overall the pipeline looks like this:

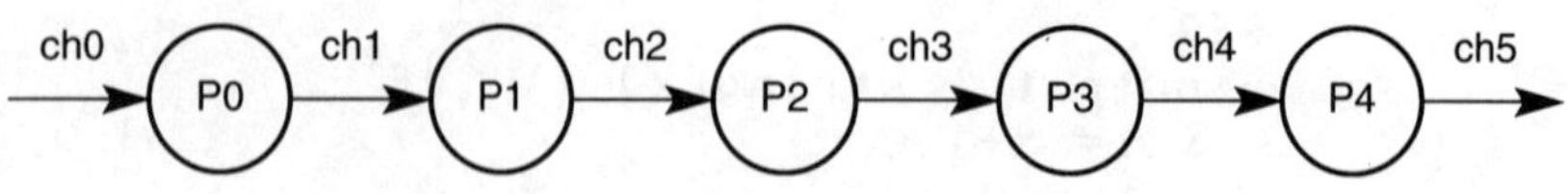

where P refers to process and ch to channel.

Consider an example program to sort 200 numbers. We can set up the pipeline by defining an array of channels:

```
[201]CHAN OF INT pipe:
```

The process below will then compare values for the *i*th pipe of the pipeline:

```
SEQ
  pipe[i] ? nextnumber
  IF
    nextnumber <= largest
      pipe[i+1] ! nextnumber
    nextnumber > largest
      SEQ
        pipe[i+1] ! largest
        largest := nextnumber
```

This would have to be repeated 199 times in each process to pass through all the numbers; and in this case, as we know in advance how many times it has to execute, a replicated SEQ is preferable to a WHILE loop:

```
INT largest:
SEQ
  pipe[i] ? largest
  SEQ j=0 FOR 199
    INT nextnumber:
    SEQ
      pipe[i] ? nextnumber
      IF
        nextnumber <= largest
          pipe[i+1] ! nextnumber
        nextnumber > largest
          SEQ
            pipe[i+1] ! largest
            largest := nextnumber
  pipe[i+1] ! largest
```

This process now needs to be replicated using the index `i`, and can also be tidied up by using abbreviations for `pipe[i]` and `pipe[i+1]`:

```
PAR i = 0 FOR 200
  in IS pipe[i]:
  out IS pipe[i+1]:
  INT largest:
  SEQ
    in ? largest
    SEQ j=0 FOR 199
      INT nextnumber:
      SEQ
        in ? nextnumber
        IF
          nextnumber <= largest
            out ! nextnumber
          nextnumber > largest
            SEQ
              out ! largest
              largest := nextnumber
    out ! largest
```

We now need two further processes, one which feeds numbers into the pipeline at `pipe[0]`, and another which bleeds the sorted numbers from `pipe[200]`. Assuming that the numbers to be sorted come from a channel called `unsort`, and that the sorted results go out on a channel called `sort`, both of which connect to processes elsewhere in a larger program, we can deal with these processes as below:

```
SEQ i=0 FOR 200
  INT outofsequence:
  SEQ
    unsort ? outofsequence
    pipe[0] ! outofsequence
SEQ i=0 FOR 200
  INT sortedsequence:
  SEQ
    pipe[200] ? sortedsequence
    sorted ! sortedsequence
```

We can now put the program together, again tidying up by the use of an abbreviation for the number of values to be sorted:

```
VAL tosort IS 200:
[tosort+1]CHAN OF INT pipe:
PAR
  PAR i=0 FOR tosort
    in IS pipe[i]:
    out IS pipe[i+1]:
    INT largest:
    SEQ
      in ? largest
      SEQ j=0 FOR tosort-1
        INT nextnumber:
        SEQ
          in ? nextnumber
          IF
            nextnumber <= largest
              out ! nextnumber
            nextnumber > largest
              SEQ
                out ! largest
                largest := nextnumber
      out ! largest
  SEQ i=0 FOR tosort
    INT outofsequence:
    SEQ
      unsort ? outofsequence
      pipe[0] ! outofsequence
  SEQ i=0 FOR tosort
    INT sortedsequence:
    SEQ
      pipe[tosort] ? sortedsequence
      sorted ! sortedsequence
```

The feed, pipeline and bleed processes are combined into a PAR so that all run at once; note that synchronization is completely catered for by the language.

The program can be made more efficient if we make use of the fact that the end of the number stream reaches sorted order progressively as we move down the pipe. Therefore, the last sorted numbers may be copied straight through without any testing. At process pipe[i], i numbers are already in sorted order and tosort-i remain unsorted:

```
PAR i=0 FOR tosort
  in IS pipe[i]:
  out IS pipe[i+1]:
  SEQ
    VAL tobesorted IS tosort-i:
    INT largest:
    SEQ
      in ? largest
      SEQ j=0 FOR tobesorted-1
        INT nextnumber:
        SEQ
          in ? nextnumber
          IF
            nextnumber <= largest
              out ! nextnumber
            nextnumber > largest
              SEQ
                out ! largest
                largest := nextnumber
      out ! largest

    VAL processed IS i:
    SEQ j=0 FOR processed
      INT straight.through:
      SEQ
        in ? straight.through
        out ! straight.through
```

In this sorting example, the number of values to be sorted is known in advance, so the feed and bleed processes could be programmed with a self-terminating replicated SEQ.

In programs where indefinite numbers of values are required to pass through a pipeline until told to stop, the feed and bleed processes will need to use WHILE loops. This leads to the general problem in concurrent programming of terminating gracefully.

For the above pipeline example, the best solution is for the feed process to send a message, through the pipeline itself, which terminates the bleed process and ensures that the pipeline is 'flushed out'. Merely terminating the feed process on exhaustion of the input stream would cause the pipeline to deadlock while still holding needed results. The use of a variant protocol on the channel would allow graceful termination.

2.9 Configuration

In most computer languages, input/output (I/O) is an afterthought as the language has been designed to manipulate data only inside the computer's memory. I/O therefore tends to be tied to specific hardware devices. With occam this is not the case; I/O is performed over channels, and all we need to know further is how to attach occam channels to real-world devices, which can then be treated as further occam processes.

In any particular occam implementation there will be a number of I/O channels. These channels may in some cases lead directly to the hardware, via driver programs which the occam compiler links to the program. In other cases they may lead to the operating system of the host computer, which then handles I/O on behalf of occam.

A typical occam system will support at least channels for a VDU screen, a keyboard and a filing system. These channels are given numbers which can be found in the manual for the particular occam implementation. If, for example, channels are numbered as:

1	Output to screen
2	Input from keyboard

these numbers can be associated with the channel names used in an occam program as follows:

```
PLACE screen AT 1:
PLACE keyboard AT 2:
```

Inside a program these channels can all be used for input and output in the usual fashion:

```
VAL string IS "Good Morning":
SEQ i=0 FOR SIZE string
  screen ! INT string[i]
```

would display the message on a VDU screen, and:

```
keyboard ? Z
```

would input the code for a single character typed at the keyboard into variable `Z`.

2.9.1 Device protocols

Some hardware devices will require various protocols to be adhered to, that is certain special commands and non-data characters may need to be sent (and received) to control the devices, in addition to the actual data.

These are very hardware dependent issues, but consider a very simple example of a buffered screen channel. Characters are only displayed when the buffer empties, happening automatically when 80 characters have accumulated or on demand if the endrecord code (for example –2) is received. Hence:

```
VAL endrecord IS -2:
...
SEQ
  VAL string IS "Good Morning":
  SEQ i=0 FOR SIZE string
    screen ! INT string[i]
  screen ! endrecord
```

Ports

In addition to hard channels, occam can address I/O **ports**. A port declaration has a data type. For example:

```
PORT OF BYTE serialcom1:
```

and the allowed processes are input and output only:

```
serialcom1 ! 'K'
serialcom2 ? Z
```

This allows ports to be used like channels rather than like variables. Therefore, as with an occam channel, only one process may input from a port and only one process may output to a port.

Note that there is no synchronization mechanism associated with port input and output. Any timing constraints which result from the use of asychronous external hardware will have to be programmed explicitly.

During applications development it is recommended that the peripheral is modelled by an occam process connected via channels.

2.9.2 Allocation

Occam programs may be designed, written, tested and debugged on a single processor workstation, and then transferred to a network of parallel computers. The final stage of such a development cycle is to allocate parallel processes in the program to different processors. This allocation is performed by replacing PAR with PLACED PAR in the appropriate parts of the program. PLACED PAR is followed by a placement, which consists of the number of a processor and a process to be run on it.

For example, the pipelined sorting program could be allocated to a linear array of transputers by the following placement:

```
PLACED PAR
  PLACED PAR i=0 FOR tosort
    PROCESSOR i
      in IS pipe[i]:
      out IS pipe[i+1]:
      INT largest:
      SEQ
        in ? largest
        SEQ j=0 FOR tosort-1
          INT nextnumber:
          SEQ
            in ? nextnumber
            IF
              nextnumber <= largest
                out ! nextnumber
              nextnumber > largest
                SEQ
                  out ! largest
                largest := nextnumber
        out ! largest
...rest of routine
...
...
```

None of these operations in any way affect the logic of a program. Therefore, a program can be developed with little or no thought to such matters, and the required configuration can be performed when the program logic is proven.

2.10 Programming style

In conclusion, a few points on programming style are necessary. A program should be factorized to a degree where the majority of the code is contained in the PROCS. The aim is a main program free from distracting detail, with a suitably named high-level PROCS. Note that this program could also be more easily altered.

Use named abbreviations freely. These abbreviations have no overhead and may improve performance by allowing the compiler to choose the most efficient implementation. The use of named constants also makes a program more understandable and readable. Names may be as long as necessary; it costs nothing in program efficiency, and for the sake of readability is well worth the extra typing.

Keep all specifications local, normally immediately before the process in which they are used. Often variables are used solely as working stores to transfer a value from an input process to an output, and are normally declared just before the process in which the input and output occur. Only when they need to be free variables with respect to some loop need they be declared earlier.

Avoid preoccupation with configuration and priority until the very end of the program development. Finalize the program logic using small stand-in procedures to emulate absent devices.

Develop the skill of detecting inherent parallelism in problems. Attempts to parallelize an existing sequential algorithm do not normally lead to the most efficient solution – a fresh analysis of the problem is essential. Factorize the problem using as much parallelism as possible; remember that virtually the same program can be executed on just a few processors or on a massively parallel system.

2.11 Future directions

Inmos have made it very clear that they fully intend to support the occam programming language. The language itself is becoming increasingly popular as the standard for the teaching of parallel processing. The properties highlighted in this chapter render it a favoured language for a growing spectrum of user applications.

Inmos have announced that an occam 2 toolset will be made available for the T9000, which promises to be source compatible with the occam 2 system in the current toolset.

The first phase of the development towards occam 3 is underway. It has been announced that this release plans to extend occam 2 by supporting user-defined types, records, arithmetic on BYTE types, the return of fixed length arrays from FUNCTIONS, the dropping of FROM and FOR in segments of arrays and the removal of the requirement for explicitly typing literals. A MODULE concept will enable cleaner design and the ability to build up a library of standard routines, each separately compiled, with a secure application interface. Occam 3 is intended to allow the sharing of channels to facilitate the multiple-client/single-server paradigm.

This represents a very positive step towards a much longer term development of the language which will involve the provision of a new compiler based on ANDF technology. This phase of development intends to address modularity and sharing, and should result in the provision of native occam 3 compilers for a wide range of different computer architectures. It has been strongly recommended that the front-end of the existing compiler be reused without alteration. This will save time, ensure correct and efficient application of full semantic checking, and will also enable re-targeted compilers to keep track of future developments in the primary transputer compiler.

Inmos are actively encouraging the re-targeting of the language to non-transputer architectures, even though occam is a registered trademark of their group of companies. Source code and documentation for the Inmos occam compiler are available as a product on a QIC-24 cartridge tape.

References

Inmos Ltd. (1988). *Transputer Architecture*. Transputer Reference Manual

Kernighan, B. W. and Richie, D. (1978). *The C Programming Language*. Prentice-Hall.

May, D., Sheperd, R. and Keane, C. (1988). *Communicating Process Architecture*. Inmos Ltd.

Zenith, S. E. (1987). *The Philosophy of Occam*. Inmos Ltd.

Chapter 3
Transputer Software

R.M. McKeag

3.1 The transputer and its software

It is still the case that the immediate appeal of a computer system is generally expressed in terms of its hardware: megabytes, MIPS and megaflops are often the first things one is told about. In practice, for most users of computers, the really important aspect is the ease with which existing and new software can be developed for and run on the computer.

The transputer is unusual in that ease of parallel programming came first; this led to the CSP notation, and this in turn led to the complementary hardware (the transputer) and software (the occam language). Indeed, occam was widely available before the transputer was on sale and so, through pencil and paper exercises and simulations on other computers, quite a lot of software had already been developed for transputer systems by the time the hardware first reached the users.

C.A.R. Hoare's pioneering work on high-level languages for parallel processing took place in Northern Ireland in the 1970s when he was Professor of Computer Science at the Queen's University of Belfast. This work, which was influenced by many discussions over the years with E.W. Dijkstra, P. Brinch Hansen and others, grew out of an examination of operating systems, their techniques, their structure, their use of parallelism and how to program them. It extended into the general field of parallel programming and language features for expressing concurrency safely. It led to the development of Pascal Plus, an extension of N. Wirth's Pascal supporting processes and monitors (or **envelopes** – class-like structures derived from O.-J. Dahl's Simula 67) and permitting very fast one pass compilation. It culminated in the design of the Communicating Sequential Process (CSP) notation, which provided a compact, simple and expressive way of describing a parallel system.

This research having been completed and CSP having been studied, used and implemented by many people the world over, Inmos was able to base the design of the transputer and occam on CSP with considerable confidence.

Once the transputer became available, it found a pool of programmers already trained, enthusiastic and ready to program it in occam. Thus a variety of new applications was quickly developed, and much practical research was undertaken into the effective use of parallel processing. Occam thus satisfied one part of the potential market for the machine.

However, other and much larger potential markets existed. These were characterized by users who were unwilling, at least initially, to use such an unfamiliar language as occam. Their reasons were perfectly valid. In some cases they already had sequential programs that they wished to speed up by taking advantage of the transputer: either they wished to run them sequentially on a single processor – in other words, they saw the transputer as a cheap and very fast coprocessor for their desktop computers – or they wished to make use of the parallelism easily obtained by connecting several transputers together. In other cases they wished to develop new software but were wary of diverting too much effort and time into getting to grips with a new way of programming (parallelism), new hardware (the transputer) and a new language (occam).

In response to this, software houses developed compilers for parallel versions of familiar languages such as C, Fortran and Pascal, with further languages including Ada, Basic, C++, Modula, Prolog and others following later. This chapter is primarily concerned with showing the reader how a sequential program in one of these languages can be readily run on a single transputer, how the program might be parallelized to run as several processes on a single transputer, and how it might then be run on a network of transputers to achieve true concurrency. The aim of the chapter is thus to address the problems of the users described in the previous paragraph, and thereby to encourage potential users of the transputer to become actual users with as little difficulty as possible. The material has been used successfully in courses for people with widely varying technical backgrounds and with experience of a great variety of application areas and programming languages; they have found that getting to grips with a new way of programming and new hardware need not be daunting, and can be accomplished in a couple of days. Often they are then keen to learn and even adopt occam as they begin to appreciate that, with its inbuilt concurrency, it has advantages over the parallel versions of sequential languages.

3.2 Porting sequential programs

The first conventional languages to be implemented on the transputer were C, Fortran and Pascal, as it was felt that the bulk of programs that were likely to be transferred to transputers were in those languages, in particular C, which was implemented by several companies, and Fortran, which was implemented by rather fewer. These are all standard sequential languages and so, in each case, an attempt was made to minimize the effect of adding parallelism: this has usually taken the form of writing each process as a separate program and providing a library of routines (procedures, functions) to enable a process to write messages to, or read messages from, other processes.

Since many companies have implemented these languages, each with its own slightly different conventions for handling interprocess synchronization, a choice must be made here to adopt one approach as illustrative of the genre. Probably the most widely used implementations are those from the Edinburgh software house 3L, which produced some of the earliest versions, namely Parallel C, Parallel Fortran and Parallel Pascal. All of these are compatible to the extent that a program can consist of, say, three processes, one in each of the three languages. Thus, in this chapter, we shall use these three compilers when discussing details or developing examples. Also, we shall assume that the computing environment in which the transputer program is to be developed and run consists of an IBM PC (or equivalent), running the MicroSoft MS-DOS operating system (or equivalent) and a text editor, and supplemented by a board containing one or more T414 or T800 transputers; this is probably the most readily available and widely used transputer configuration.

Readers who have access to different systems should be able to understand the principles of operation expounded in this chapter and then, with some persistence, apply them elsewhere. The details of the language and subroutine library may be different; the documentation may be different (on the whole, the 3L documentation is clear); the operating system and the commands for editing, compiling, linking, configuring and running programs may be different; the editor may be different, and the method of configuring the transputers and the processes running on them may be different. And, if you find the exercise frustrating, remember that most of your problems probably emanate not from the transputers and their parallelism, but from the complex layers of proprietary software heaped onto them by a well-meaning but misguided vendor of system software.

For many users who have existing software, the easiest introduction to using transputers is to recompile an existing program to run as a single process on a single transputer; very often a single transputer will give a worthwhile improvement in performance over the host computer (for example, a PC). This resembles porting software to any other machine. As usual, one has to check the existence of a suitable compiler, its implementation-dependent features, perhaps the speed of the generated code, the availability of suitable libraries and access to peripheral devices. Provided one has used only standard features, there should be little or no difficulty in recompiling one's program. The usual cause of problems is that one's existing program uses non-standard features: perhaps the program is in a pre-standard version of the language (a common C problem); perhaps it uses features peculiar to the manufacturer (a common Fortran problem); perhaps it comes from a system that has an unusual extension for including program files or rewinding data files (common Pascal problems).

When one's program is loaded into a transputer a server program is loaded into the host computer and through this the host's peripheral devices may be used by the program in the transputer. Provided the only devices one needs to use are the host computer's keyboard and (text) screen and (occasionally) disk files, there should be few, if any, problems. However, if one wishes to use some exotic device it may be necessary to rewrite the server program – the control of a robot arm is an example. To output to a high resolution graphics screen or to enjoy fast access to files, it will probably be necessary to drive the device directly, without using the software in the host as an intermediary – this necessitates a board and software for driving the device from the transputer: your system may have libraries of routines for doing this.

Once your program is stored in a file on the host computer you will need to invoke, in turn, the compiler, the linker and the loader. To compile your program from a file, say **prog** (with a **.c** or **.f77** or **.pas** extension), to produce a file called **prog.bin** (suitable for running on a T414 or a T800 transputer), type one of the following:

t4c prog	or	**t8c prog**	if your program is in **C**,
t4f prog	or	**t8f prog**	if your program is in **Fortran**,
t4p prog	or	**t8p prog**	if your program is in **Pascal**.

To link the binary object file, say **prog.bin**, with the run-time library to produce an executable program called **prog.b4** (whether it is to run on a T414 or a T800) type one of the following:

t4clink prog	or	**t8clink prog**	if your program is in **C**,
t4flink prog	or	**t8flink prog**	if your program is in **Fortran**,
t4plink prog	or	**t8plink prog**	if your program is in **Pascal**.

Executable programs such as **prog.b4** are loaded into the transputer and run using the **afserver** program, which is invoked by typing:

afserver –:b prog.b4 –:o 1

Omitting the **–:o 1** parameter is a frequent cause of program failure as programs run in this way are limited in their stack usage to the transputer's very fast but small (2 Kbyte) on-chip stack; some Fortran and almost all C and Pascal programs require more than this, and the parameter signifies that the program is to use an off-chip stack. Considerable improvements in performance can be obtained by sensible use of the transputer's on-chip memory (2 Kbyte on the T414 and 4 Kbyte on the T800), and facilities exist, when setting up the configuration, to place the stack or some of the most frequently used or time-critical routines on the chip. The **afserver** command may be followed by further parameters which can be made available to your program. The **afserver** does two things: first, it loads your program onto the transputer network and starts it running; second, it waits for and acts upon calls from the nearest (or **root**) transputer to read from the host's keyboard, write to its screen or use its disk.

This simple sequence of commands which, if used frequently, can be defined in a macro, is usually all that is needed to run your program on a single transputer.

If you have a transputer system to hand it is worthwhile running a sequential program in this way, if only to boost your confidence before tackling parallelism. The example given here is of a program that, without performing any checks on its input, reads 20 integers from the host computer's keyboard and echoes them back, negated, to the host's screen, one per line; this simple program will form the basis of the examples used later in the chapter.

In C it takes the form:

```
/* prog1.c */
#include <stdio.h>
    main( )
    {
        #define limit 20
        int  number[limit], count;
        printf("Please type 20 integers:\n");
        for (count = 0; count < limit; ++count) scanf("%i", &number[count]);
        for (count = 0; count < limit; ++count) printf('%i\n", –number[count]);
    }
```

and the necessary commands for compiling and running it on a T800 are:

t8c prog1
t8clink prog1
afserver –:b prog1.b4 –:o 1

In Fortran it takes the form:

```
C      prog1.f77
       PROGRAM prog1
       PARAMETER (limit = 20)
       INTEGER number(limit), count
       PRINT *, 'Please type 20 integers:'
       READ *, (number(count), count = 1, limit)
       DO 100 count = 1, limit
               PRINT *, –number(count)
  100  CONTINUE
       END
```

and the necessary commands are:

t8f prog1
t8flink prog1
afserver –:b prog1.b4

In Pascal it takes the form:

```
{prog1.pas}
program prog1 (input, output);
        const
                limit = 20;
        var
                number: array [1..limit] of integer;
                count: 1..limit;
        begin
                writeln('Please type 20 integers:');
                for count := 1 to limit do read(number[count]);
                for count := 1 to limit do writeln(–number[count])
        end.
```

and the necessary commands are:

t8p prog1
t8plink prog1
afserver –:b prog1.b4 –:o 1

3.3 Specifying the configuration

The **prog1** example program used in the last section was run as a single process (or **task** in 3L parlance) on a single transputer connected to a host computer, which ran the **afserver** program. In addition, a small **filter** process was run on the transputer between the **afserver** and **prog1**. The **filter** is required because some versions of the transputer cannot transmit one byte messages, which thus have to be padded out to two bytes. So even such a simple program involved two processors and three processes (Figure 3.1). In this case, there was no need to configure the system explicitly as the normal linker **t4Xlink** or **t8Xlink** (where **X** denotes **c**, **f** or **p**, depending upon the language used) links in with the user's program the runtime library and a 'harness' which consists of the **filter** process and some system initialization code.

In general, however, one must write a configuration file, say **prog1.cfg**, and invoke both a different linker **t4Xtask** or **t8Xlink** (where **X** denotes **c**, **f** or **p**) and a configurer **config**. This new linker links in with the user's program

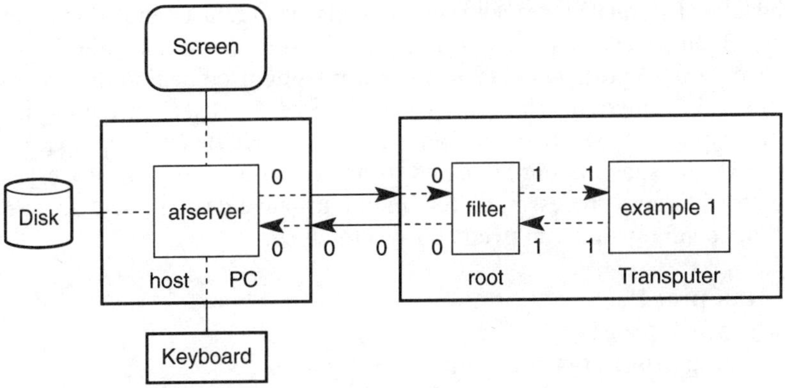

Figure 3.1 A configuration of three tasks on two processors.

the runtime library and a vestigial harness with neither **filter** nor system initialization code. To illustrate this we can explicitly configure our **prog1** system by writing the following configuration file:

```
! prog1.cfg
! configure the processors by specifying how their links are wired:
processor  host              ! the PC
processor  root              ! the transputer
wire  ?  host[0] root[0]     ! wire <name> <link> <link>
! configure the tasks by specifying how their input and output ports are
connected:
task  afserver  ins = 1 outs = 1       ! task <name> <inputs>
<outputs> <data>
task  filter    ins = 2 outs = 2 data = 10k
task  prog1     ins = 2 outs = 2
connect  ?  afserver[0] filter[0]      ! connect <name> <output>
<input>
connect  ?  filter[0] afserver[0]
connect  ?  filter[1] prog1[1]
connect  ?  prog1[1]  filter[1]
! configure the system by specifying how the tasks are to be placed on
the processors:
place afserver  host          ! place <task name> <processor name>
place filter  root
place prog1  root
```

As can be seen, this describes the physical configuration of the processors, the logical configuration of the processes and the mapping between the two. The wires connecting pairs of processors and the channels connecting pairs of ports can be named or, as they are not referred to elsewhere, just denoted by question marks. The **afserver** has one pair of ports; the runtime library used by **prog1** uses two pairs of ports, the pair numbered **0** being used behind the scenes for diagnostic information, while the pair numbered **1** is used for data transfers. The **filter** process uses two pairs and requires 10 Kbytes of workspace, with the rest of the space being allocated to **prog1**. Where several processes are placed on the same processor, all (but one) must have data clauses. So, to compile, link, configure and run **prog1** on a T800, type:

t8X prog1
t8Xtask prog1
config prog1.cfg prog1.app
afserver –:b prog1.app –:o 1

replacing **X** by **c**, **f** or **p**. As mentioned earlier, this explicit configuration is unnecessary where **prog1** consists of a single process, but it illustrates well how to handle programs consisting of multiple processes – the subject of the next section of this chapter.

Other software systems use different conventions for configuring transputer networks, but in each case the wiring of the processors' links, the connection of the processes' input and output ports by logical channels and the placement of processes on processors must be explicitly or implicitly specified.

The wiring of the processors' links may take three forms, depending upon the hardware. Links may be permanently wired together, for example link 0 of the transputer on a B004 board to link 0 of the PC, or the four transputers on a B003 board are connected by their links 2 and 3 to form a square. Links may be physically connected by short lengths of wire, for example links 0 and 1 of each of the four transputers on a B003 board may be connected to transputers on the same board or other boards by hand. Links may be connected by an electronic switch, in which case a program must be run to set up the switch – in this case it is not normal to reconfigure the system while it is running. Obviously, the degree of flexibility of the target configuration can affect the way in which the programmer designs and configures the program.

In some systems, processes may communicate with each other only if they are on the same transputer or on physically connected transputers; in other systems, two distant processes may communicate, either through an electronic switch or through the agency of small routing processes placed by the system in each processor. Again, knowledge of this can affect program design and configuration.

Some systems automate the placement of processes on processors to a certain extent, but for load balancing the user usually wishes to retain control over this. Dynamically changing the placement is not generally possible; it may perhaps best be achieved by explicitly placing copies of a process on several transputers and then deciding at runtime, perhaps for load balancing purposes, whither to send the packets of work; this is a variant of a **processor farm**, which is the topic of a later section of this chapter.

3.4 Programming with multiple processes

A program consisting of multiple processes (possibly in different languages) is organized so that one of the processes (called the driver) is placed on the root transputer and communicates with the **afserver** in the host PC, while the other processes are spread over the network of transputers. In the 3L systems, normally only the driver process can use the standard input and output operations for communicating (via the **filter**) with the **afserver**; however, 3L provides a multiplexer process **filemux** which can be interposed between the filter and several driver processes. Some other systems allow any process to use such operations, perhaps to display messages on its own window on the host's screen; care must be taken to ensure that only one process at a time uses the keyboard (perhaps the process whose window is currently selected) or the disk drive. Indeed, it is often simpler and safer to channel communication explicitly through the driver unless one has a mastery of the principles of operating system synchronization and mutual exclusion and, of course, the mechanisms to implement them conveniently. Despite the wealth of knowledge that has been gleaned during the past 20 years about high-level parallel programming languages, most transputer programming systems, being based on sequential languages, provide only primitive synchronization functions which are notoriously difficult to use safely.

To write a program consisting of multiple processes, write a program for each, possibly using different languages, and store each process in a separate file. Processes communicate with one another using a library of channel communication operations instead of the usual C, Fortran or Pascal input and output commands: these operations, which are provided in an 'include' file **chan**, enable a process to input or output a byte or a word or a message. Synchronization is effected by the system ensuring that the communication takes place only when the two processes connected by a channel have invoked a complementary pair of operations; most systems provide ways of circumventing this protocol, but such temptations should always be scorned. Each process, except for the driver, is linked with the 'standalone' version of the runtime library: this produces more compact code because it does not include the input and output functions that require **afserver** support. This linking is done using **t4Xstask** instead of **t4Xtask**, and **t8Xstask** instead of **t8Xtask**, where **X** denotes **c**, **f** or **p**.

To illustrate this, we shall extend the program **prog1** to two processes: the first, **prog2a**, reads 20 integers and passes them to the second, **prog2b**, which negates them and returns them to the first for printing. Only **prog2a** communicates with the **afserver** and is thus the driver, and we link in the full runtime library using **t8Xtask**; **prog2b** requires only the standalone runtime library and is linked with **t8Xstask**. The commands are as follows:

t8X prog2a
t8X prog2b
t8Xtask prog2a
t8Xstask prog2b
config prog2.cfg prog2.app
afserver –:b prog2.app -:o 1

Note that when compiling Pascal with non-standard extensions, as will be the case here, one has to add an **/X** qualifier thus:

t8p prog2b /X

The configuration file **prog2.cfg** takes the following form, where the two processes occupy the same transputer. This necessitates supplying a **data** clause for one or both of the processes; estimating the amount of space that might be required is difficult, although a **decode** utility can be used to measure the code and static workspace required by each process; however, the stack space and heap space can be difficult to guage and so, unless space is scarce, a wildly generous overestimate is often used:

```
! prog2.cfg
! configure the processors by specifying how their links are wired:
processor   host            ! the PC
processor   root            ! the transputer
wire  ?  host[0] root[0]    ! wire <name> <link> <link>
! configure the tasks by specifying how their input and output ports are
connected:
task  afserver  ins = 1 outs = 1      ! task <name> <inputs>
<outputs> <data>
task  filter    ins = 2 outs = 2  data = 10k
task  prog2a    ins = 3 outs = 3  data = 500k
task  prog2b    ins = 2 outs = 2
connect  ?  afserver[0] filter[0]     ! connect <name> <output>
<input>
connect  ?  filter[0] afserver[0]
connect  ?  filter[1] prog2a[1]
connect  ?  prog2a[1] filter[1]
connect  ?  prog2a[2] prog2b[1]
connect  ?  prog2b[1] prog2a[2]
! configure the system by specifying how the tasks are to be placed on
the processors:
place afserver  host  ! place <task name> <processor name>
place filter    root
place prog2a root
place prog2b root
```

If the same program is to be run using two transputers, the processes and their linking remain unaltered but the configuration file must be amended as follows:

```
processor        branch      ! the second transputer
wire   ?  root[1] branch[0]  ! wiring the two transputers together
place  prog2b    branch      ! replaces the last line of the file
```

To illustrate the use of the channel communication operations the processes will be programmed in each of the three languages we are using; the reader might try to run these in various combinations using either one or two transputers.

In C the header file **chan.h** is provided, and it includes the channel data type **CHAN** and the three pairs of operations with which we are concerned here:

```
void chan_in_byte (byte, channel)
void chan_in_word (word, channel)
void chan_in_message(length, message, channel)
```

where:

```
int length, *word;
char *byte, *message;
CHAN *channel;
```

and:

```
void chan_out_byte (byte, channel)
void chan_out_word (word, channel)
void chan_in_message(length, message, channel)
```

where:

```
int length, word;
char byte, *message;
CHAN *channel;
```

The relationship between each channel and the ports it connects has already been specified in the configuration file:

```
task     prog2a       ins = 3 outs = 3      data = 500k
task     prog2b       ins = 2 outs = 2
connect      ?        prog2a[2]    prog2b[1]
connect      ?        prog2b[1]    prog2a[2]
```

and this information is transmitted to the process through its argument list:

```
main (argc, argv, envp, in_ports, ins, out_ports, outs)
int argc, ins, outs;
char *argv[], *envp[];
CHAN *in_ports[], *out_ports[];
```

Here **argc** denotes the number, and **argv** the values, of the tokens or arguments following **afserver** and its own arguments (**–:b**, **–:o**) on the command line, except that **argv[0]** is always a pointer to a null string instead of the program name; thus **argv[1]** will be a pointer to the string **'datafile'** if the command line reads:

afserver –:b prog2.app –:o 1 datafile

and **envp** is always null.

The **ins** and **outs** arguments tell the program how many channels there are; **in_ports** and **out_ports** are vectors of pointers to the channels; thus, to send an integer from **prog2a** to **prog2b** the former would use:

chan_out_word(intvalue, out_ports[2])

and the latter would use:

chan_in_word(intvariable, in_ports[1])

We can now program process **prog2a.c**:

```
/* prog2a.c */
#include <stdio.h>
#include <chan.h>
main (argc, argv, envp, in_ports, ins, out_ports, outs)
int argc, ins, outs;
char *argv[ ], *envp[ ];
CHAN *in_ports[ ], *out_ports[ ];
{
    #define limit 20
    int number[limit], count;
    printf("Please type 20 integers:\n");
    for (count = 0; count < limit; ++count) scanf("%i", &number[count]);
    for (count = 0; count < limit; ++count)
    {
        chan_out_word(number[count], out_ports[2]);
        chan_in_word(&number[count], in_ports[2]);
        printf("%i\n", number[count]);
    }
}
```

and process **prog2b.c**:

```
/* prog2b.c */
#include <chan.h>
main (argc, argv, envp, in_ports, ins, out_ports, outs)
int argc, ins, outs;
```

```
char *argv[], *envp[];
CHAN *in_ports[], *out_ports[];
{
    #define limit 20
    int  number, count;
    for (count = 0; count < limit; ++count)
    {
        chan_in_word(&number, in_ports[1]);
        chan_out_word(–number, out_ports[1]);
    }
}
```

In Fortran the 'include' file **CHAN.INC** provides the three pairs of operations:

```
F77_CHAN_IN_BYTE (BYTE, CHANNEL)
F77_CHAN_IN_WORD (WORD, CHANNEL)
F77_CHAN_IN_MESSAGE(LENGTH, MESSAGE, CHANNEL)
```

and:

```
F77_CHAN_OUT_BYTE (BYTE, CHANNEL)
F77_CHAN_OUT_WORD (WORD, CHANNEL)
F77_CHAN_OUT_MESSAGE(LENGTH, MESSAGE, CHANNEL)
```

where:

```
INTEGER BYTE, LENGTH, CHANNEL
```

and **WORD** may be **INTEGER**, **REAL** or **CHARACTER*4** and **MESSAGE** may be any variable, scalar or array, of at least **LENGTH** bytes.

Also provided are two vectors of ports with one array element representing each channel:

```
INTEGER F77_CHAN_IN_PORT
INTEGER F77_CHAN_OUT_PORT
```

so that, given:

```
connect  ?  prog2a[2]  prog2b[1]
```

in the configuration file, the processes **prog2a** and **prog2b** can communicate by the former invoking:

```
CALL F77_CHAN_OUT_WORD(INTVALUE, F77_CHAN_OUT_PORT(2))
```

while the latter invokes:

```
CALL F77_CHAN_IN_WORD(INTVARIABLE, F77_CHAN_OUT_ PORT(1))
```

We can now write process **prog2a.f77**:

```
C           prog2a.f77
            PROGRAM prog2a
            INCLUDE 'CHAN.INC'
            PARAMETER (limit = 20)
            INTEGER number(limit), count
            PRINT *, 'Please type 20 integers:'
            READ *, (number(count), count = 1, limit)
            DO 100 count = 1, limit
                    CALL F77_CHAN_OUT_WORD(number(count),
                                             F77_ CHAN_ OUT_ PORT (2))
                    CALL F77_CHAN_IN_WORD(number(count),
                                              F77_ CHAN_ IN_ PORT (2))
                    PRINT *, number(count)
100         CONTINUE
            END
```

and process **prog2b.f77**:

```
C           prog2b.f77
            PROGRAM prog2b
            INCLUDE 'CHAN.INC'
            PARAMETER (limit = 20)
            INTEGER number, count
            DO 100 count = 1, limit
                    CALL F77_CHAN_IN_WORD(number,
                                              F77_CHAN_ IN_ PORT (1))
                    CALL F77_CHAN_OUT_WORD(–number,
                                             F77_CHAN_ OUT_ PORT(1))
100         CONTINUE
            END
```

In Pascal the 'include' file **chan.inc** provides the channel data type **Chan_Type** and the three pairs of operations:

```
procedure Chan_In_Byte (var byte: UNIV char; var channel: Chan_ Type)
procedure Chan_In_Word (var word: UNIV integer; var channel: Chan_Type)
procedure Chan_In_Message
        (length: integer; var message: UNIV char; var channel:Chan_Type)
```

and:

```
procedure Chan_Out_Byte (byte: char; var channel: Chan_Type)
procedure Chan_Out_Word (word: integer; var channel: Chan_Type)
procedure Chan_Out_Message
        (length: integer; var message: UNIV char; var channel: Chan_Type)
```

where **UNIV**, when prefixing a **var** parameter, denotes that the actual parameter can be of any type – this is, of course, wildly agin the spirit of Pascal.

Also provided are the integer variables **In_Ports** and **Out_Ports** in which are held the numbers of input and output channels associated with the process's ports. The channels are accessed through **In_Chan** and **Out_Chan** which are pointers to the two vectors of ports. Each port is a pointer to a channel variable. Thus, given:

```
connect ? prog2a[2] prog2b[1]
```

in the configuration file, **prog2a** can execute:

```
Chan_Out_Word(intvalue, Out_Chan^[2]^)
```

while **prog2b** executes:

```
Chan_In_Word(intvariable, In_Chan^[1]^)
```

to pass an integer from one to the other.

We can now program process **prog2a.pas**:

```
{prog2a.pas}
program prog2a (input, output);
    $include 'chan.inc'
    const
        limit = 20;
    var
        number: array [1..limit] of integer;
        count: 1..limit;
    begin
        writeln('Please type 20 integers:');
        for count := 1 to limit do read(number[count]);
        for count := 1 to limit do
        begin
            Chan_Out_Word(number[count], Out_Chan^[2]^);
            Chan_In_Word(number[count], In_Chan^[2]^);
            writeln(number[count])
        end
    end.
```

and **prog2b.pas**, which illustrates a peculiar feature of 3L Pascal: in each process other than the driver the heading begins with **module** instead of **program** and, at the outermost level, variable declarations begin with **static** instead of **var**:

```
{prog2b.pas}
    module prog2b;
        $include 'chan.inc'
        const
            limit = 20;
        static number: integer;
        static count: 1..limit;
        begin
            for count := 1 to limit do
            begin
                Chan_In_Word(number, In_Chan^[1]^);
                Chan_Out_Word(–number, Out_Chan^[1]^)
            end
        end.
```

As an illustration of parallel processing, **prog2** is rather poor as the two processes are never both working simultaneously; the reader might like to construct a more effective example.

3.5 Writing schedulers, multiplexers and collators

When working with mutiple processes it is often the case that a server process needs to input messages from several other processes in the order that the messages become available, although it does not know *a priori* what that order will be; one example of this is a scheduler that allocates permission to use a particular device or to write to a particular disk file or to read from the host's keyboard; a similar example is a multiplexer that enables several interprocess channels to be mapped onto a single interprocessor link; a different example is a master process that is collating packets of results generated by a number of worker processes.

The main characteristic is that performing an input operation on a named channel is inadequate as the server process does not know which process will require its services next. Thus an 'include' file provides two similar operations to allow the server to wait for the first of a number of channels.

In C the 'include' file **alt.h** provides:

```
int alt_wait (number, channel0, ..., channelnumberminus1)
int alt_wait_vec (number, channels)
```

where:

```
int number;
CHAN *channel0, ..., *channelnumberminus1, *channels[number];
```

Thus the result of calling **alt_wait(2, channel0, channel1)** will be **0**, in which case we input a message from **channel0**, or **1**, in which case we input a message from **channel1**. The alternative form uses a vector of pointers to the channels.

In Fortran the 'include' file **ALT.INC** provides:

```
INTEGER F77_ALT_WAIT (number, channel1, ..., channelnumber)
INTEGER F77_ALT_WAIT_VEC (number, channels)
```

where:

```
INTEGER number, channel1, ..., channelnumber, channels(number)
```

Thus the result of calling **F77_ALT_WAIT(2, channel1, channel2)** will be **1**, in which case we input a message from **channel1**, or **2**, in which case we input a message from **channel2**. The alternative form uses a vector of channel word addresses.

In Pascal the 'include' file **alt.inc** provides seven functions **Alt_Wait_1** .. **Alt_Wait_6** and **Alt_Wait_Vec**:

```
        function Alt_Wait_number (number, channel0, ...,
channelnumberminus1): integer
        function Alt_Wait_Vec (number, channels): integer
```

where:

```
number: integer;
channel0, ..., channelnumberminus1: Chan_Type;
channels: array [0..numberminus1] of Chan_Type;
```

Thus the result of calling **Alt_Wait_2(2, channel0, channel1)** will be **0**, in which case we input a message from **channel0**, or **1**, in which case we input a message from **channel1**. The alternative form uses a vector of pointers to the channels.

In the next section we shall look at the specialized features provided for using a processor 'farm' for distributing the workload over a network of transputers in a balanced way.

3.6 Using processor farms

In many applications it is convenient to have a **master** process that farms out packets of work to several identical **worker** processes which, independently, compute the results and return them to the master.

The 3L software allows the user to program this without the need to know, even at runtime, how many worker processes there are and where they are located. The flood-fill configurer will automatically place a worker on each transputer in the network and the master on the root transputer, and it will look after the runtime scheduling of workers and the routing of messages between the master and the workers by placing a small (11 Kbyte) process **frouter** on each transputer.

Each worker uses the library routine **net_receive** to receive a work packet, and it sends its results back using **net_send**. It cannot communicate with the host (using operations such as C's **printf**, Fortran's **PRINT** * or Pascal's **writeln**) or with other transputer processes using operations such as 3L's **chan_out_message**.

The master process generates the packets of work and sends them to the anonymous workers using **net_send**; it also receives the results using **net_receive**. The way in which the master multiplexes itself among these operations is by using multiple 'threads': thus it may use the operation **thread_create** to run a subroutine (function, procedure) as a subprocess in quasi-parallel with itself. Thus the structure of each process takes the following form (expressed in no particular language):

```
master:  {thread_create(SEND); RECEIVE}
```

where **SEND** is:

```
repeatedly {produce workpacket; net_send(workpacket)}
```

and **RECEIVE** is:

```
repeatedly {net_receive(resultpacket); consume resultpacket}
```

and:

```
worker:    repeatedly
           {
                net_receive(workpacket);
                process workpacket to produce resultpacket;
                net_send(resultpacket)
           }
```

In the master the subprocess **SEND** runs concurrently with the main process, which is an invocation of **RECEIVE**; true concurrency is not possible because both 'threads' run on the same transputer. As packets may take different times to process, perhaps because of the data they contain or because they are processed on transputers of different speeds, it is advisable to transmit a tag identifying the data, both in the work packet and in the result packet, so that the results can be assembled in the right order.

Note that the main thread of the process must not terminate until all the other threads have finished their useful work: this is the case here, where everything that is sent by **SEND** must subsequently be received back by **RECEIVE**. However, if both **SEND** and **RECEIVE** were run as new threads, the main thread would have to wait until they (or **RECEIVE** at any rate) had finished; this is best accomplished by declaring a channel **finished**, and having the main thread input from it as many times as there are subsidiary threads outputting to it as they finish, thus:

master: **var finished: chan_type; chan_init(finished);**
thread_create(SEND); thread_create(RECEIVE);
for i := 1 to 2 do chan_in_byte(dummy, finished)

where **SEND** is:

repeatedly {produce **workpacket; net_send(workpacket)**}
chan_out_byte(dummy, finished)

and **RECEIVE** is:

repeatedly {net_receive(resultpacket); consume **result packet}**
chan_out_byte(dummy, finished)

Because both **SEND** and **RECEIVE** have access to the global variables of the master, there is considerable scope for mistakes to be made; this method of parallelism is thoroughly unsafe as it relies on the programmer to guarantee mutual exclusion, that is to ensure that only one thread at a time uses common data or shared resources such as a disk file or screen output. In 3L, as in most systems, **semaphores** with **wait** and **signal** operations are provided to enable threads to synchronize with one another's activities. The semaphore was a brilliant invention, but is notoriously difficult to use properly. Thus this method of parallel programming was discarded in the mid-1970s and replaced by high-level language constructs, which, to a considerable extent, could be checked for safe use at compile time. Consequently, the professional programmer will generally eschew the use of semaphores, and will keep each thread as independent as possible, reading from and, certainly, writing to local variables only; where threads need to communicate with one another, channels between them can be declared and operations such as **chan_in_message** and **chan_out_message** can be used, as in the example of thread termination above.

To illustrate the use of processor farming with multiple threading, we shall adapt our previous example to farm out the 20 integers that are to be negated. Again, we shall use each of C, Fortran and Pascal in turn. The commands are as follows (where the **t8p** command must be qualified by **/X** as non-standard extensions are used, for example, **t8p prog3m /X**):

t8X prog3m
t8Xtask prog3m
t8X prog3w
t8Xstask prog3w
fconfig prog3.cfg prog3.app
afserver –:b prog3.app –:o 1

Note that the flood-fill configurer **fconfig** is used. The commendably brief configuration file, **prog2.cfg**, takes the following form, where one of the processes must have a data clause – often the worker is the larger as it does the bulk of the work:

```
! prog3.cfg
task master file=prog3m data=100k
task worker file=prog3w
```

C provides the 'include' file **net.h** with the pair of operations:

```
int net_send (length, packet, complete)
```

where:

```
char *packet;
int length, complete;
```

and:

```
int net_receive (packet, complete)
```

where:

```
char *packet;
int *complete;
```

The numbers of bytes sent and received are denoted by **length** and the function results. A long message must be broken into packets of at most 1024 bytes; **complete** is set to **1** for the final packet of a message and **0** otherwise. Another 'include' file, **thread.h**, provides the operation:

```
char *thread_create (routine, workspace, nargs, arg1, ..., argn)
```

where:

```
void (*routine) ();
int workspace, nargs, arg1, ..., argn;
```

This enables the function **routine** to be run concurrently as a new thread; the size of the workspace it requires must be provided, and a pointer to this workspace is returned as the result of **thread_create** for subsequent disposal when the thread has terminated, or **NULL** is returned if there is insufficient space; a number of arguments, which will be passed to **routine**, may be supplied. Should there be a need for communication between threads of the same process, an internal channel may be declared and then initialized by a call on the function:

void chan_init (chan)

of **chan.h,** where:

CHAN *chan;

It can then be used in the same way as the channels bound to the process's input and output ports.

In C the master process **prog3m.c** is:

```
/* prog3m.c */
#include <stdio.h>
#include <net.h>
#include <thread.h>
#define limit 20
#define workspace 10000

void SEND ()
{
    int complete = 1;
    struct workpacket {int index; int data;} packet;
    int count;
    printf("Please type 20 integers:\n");
    for (count = 0;count < limit; ++count)
    {
        packet.index = count;
        scanf("%i", &packet.data);
        net_send(sizeof(packet), &packet, complete);
    }
}

void RECEIVE ()
{
    int complete;
    int count;
    int number[limit];
    struct resultpacket {int index; int data;} packet;
    for (count = 0; count < limit; ++count)
    {
        net_receive(&packet, &complete);
        number[packet.index] = packet.data;
    }
    for (count = 0; count < limit; ++count) printf("%i\n", number[count]);
}

main ()
{
    thread_create(SEND, workspace, 0);
    RECEIVE();
}
```

where **SEND** and **RECEIVE** are distinct and can create no possible side-effects; the worker process **prog3w.c** takes the form of an infinite loop, as we cannot know how many times each worker will be invoked:

```
/* prog3w.c */
#include <net.h>
main ()
{
    int complete;
    struct workpacket {int index; int data;} packet;
    for (;;)
    {
        net_receive(&packet, &complete);
        packet.data = –packet.data;
        net_send(sizeof(packet), &packet, complete);
    }
}
```

In Fortran the 'include' file **NET.INC** provides the pair of operations:

```
F77_NET_SEND (LENGTH, PACKET, COMPLETE)
F77_NET_RECEIVE (LENGTH, PACKET, COMPLETE)
```

where:

```
INTEGER LENGTH
LOGICAL COMPLETE
```

and **PACKET** may be variable, scalar or array, of at least **LENGTH** bytes.

A long message must be broken into packets of at most 1024 bytes; **COMPLETE** is set to **.TRUE.** for the final packet of a message and **.FALSE.** otherwise. The 'include' file **THREAD.INC** provides the operation:

```
LOGICAL F77_THREAD_CREATE (ROUTINE, WORKSPACE, NARGS,
                                   ARG1, ..., NARGS)
```

where:

```
EXTERNAL ROUTINE
INTEGER WORKSPACE, NARGS
```

This subroutine, whose result denotes whether or not the specified amount of workspace was available, enables the subroutine **ROUTINE** to be run concurrently as a new thread; the workspace cannot subsequently be disposed of when the thread has terminated; a number of arguments, which will be passed to **ROUTINE**, may be supplied. Should there be a need for communication between threads of the same process, an internal channel may be declared and then initialized, using the 'include' file **CHAN.INC**, as follows:

```
INTEGER CHANWORD, CHANADDR
CHANADDR = F77_CHAN_ADDRESS(CHANWORD)
CALL F77_CHAN_INIT(CHANADDR)
```

CHANADDR can then be used in the same way as the channels bound to the process's input and output ports.

In Fortran the master process **prog3m.f77** is:

```
C  prog3m.f77

   SUBROUTINE SEND
   INCLUDE 'NET.INC'
   PARAMETER (limit = 20)
   PARAMETER (length = 8)
C  length, in bytes, of packet of 2 integers
   INTEGER packet(2), count, number(limit)
C  packet(1) is the index;  packet(2) is the data
   PRINT *, 'Please type 20 integers:'
   READ *, (number(count), count = 1, limit)
   DO 100 count = 1, limit
       packet(1) = count
       packet(2) = number(count)
       CALL F77_NET_SEND(length, packet, .TRUE.)
100    CONTINUE
   END

   SUBROUTINE RECEIVE
   INCLUDE 'NET.INC'
   PARAMETER (limit = 20)
   LOGICAL complete
   INTEGER packet(2), count, length, number(limit)
C  packet(1) is the index;  packet(2) is the data
   DO 100 count = 1, limit
       CALL F77_NET_RECEIVE(length, packet, complete)
       number(packet(1)) = packet(2)
100    CONTINUE
   DO 200 count = 1, limit
       PRINT *, number(count)
200    CONTINUE
   END

   PROGRAM prog3m
   INCLUDE 'THREAD.INC'
   EXTERNAL SEND, RECEIVE
```

```
      PARAMETER (workspace = 10000)
      LOGICAL threadOK
          threadOK = F77_THREAD_CREATE(SEND, workspace, 0)
          CALL RECEIVE
      END
```

and process **prog3w.f77**:

```
C     prog3w.f77
      PROGRAM prog3w
      INCLUDE 'NET.INC'
      LOGICAL complete
      INTEGER packet(2), length
C     packet(1) is the index;  packet(2) is the data
      DO 100 WHILE (.TRUE.)
          CALL F77_NET_RECEIVE(length, packet, complete)
          packet(2) = –packet(2)
          CALL F77_NET_SEND(length, packet, complete)
100       CONTINUE
      END
```

In Pascal the 'include' file **net.inc** provides the pair of operations:

```
procedure Net_Send (length: integer; var packet: UNIV char; complete:
                        Boolean)
function Net_Receive (var packet: UNIV char; var complete: Boolean):
    integer
```

where **length** and the function result denote the number of bytes sent and received. A long message must be broken into packets of at most 1024 bytes; **complete** is set to **true** for the final packet of a message, and **false** otherwise. Another 'include' file, **thread.inc**, provides the operation:

```
function Thread_Create (procedure routine; workspace, NARGS: integer):
    Boolean
```

This function, whose result denotes whether or not the specified amount of workspace was available, runs **routine** in parallel with the process that invokes it. Unlike C or Fortran, Pascal will not allow a variable number of parameters, and so variants of this function are available for calling routines that have parameters; this version requires the actual parameter corresponding to **NARGS** to be **0**. An alternative way in which to pass parameters to **routine** is to transmit them over an internal channel. Using the facilities of the 'include' file **chan.inc**, such a channel may be declared **var Channel: Chan_Type**, then initialized by a call on the procedure **Chan_Init (Channel)** and thereafter used in the same way as the channels bound to the process's ports.

In Pascal the master process **prog3m.pas** is:

```
{prog3m.pas}
program prog3m (input, output);
  $include 'net.inc'
  $include 'thread.inc'
  const
    limit = 20;
    workspace = 10000;
  var
    threadOK: Boolean;  {denotes enough space to create a thread}

  procedure SEND;
    const
      length = 8;  {length of packet of 2 integers}
    var
      packet: record index, data: integer end;
      count: 1..limit;
    begin
      writeln('Please type 20 integers:');
      for count := 1 to limit do
      begin
        packet.index := count;
        read(packet.data);
        Net_Send(length, packet, true)
      end
    end;

  procedure RECEIVE;
    var
      packet: record index, data: integer end;
      number: array [1..limit] of integer;
      length, count: integer;
      complete: Boolean;
    begin
      for count := 1 to limit do
      begin
        length := Net_Receive(packet, complete);
        number[packet.index] := packet.data
      end;
      for count := 1 to limit do writeln(number[count])
    end;

  begin
    threadOK := Thread_Create(SEND, workspace, 0);
    RECEIVE
  end.
```

and **prog3w.pas** is:

```
{prog3w.pas}
module prog3w;
    $include 'net.inc'
    static packet: record index, data: integer end;
    static complete: Boolean;
    static length: integer;
    begin
        while true do
        begin
            length := Net_Receive(packet, complete);
            packet.data := –packet.data;
            Net_Send(length, packet, complete)
        end
    end.
```

As a result of experience gained in the use of flood-filled configurations, 3L has expanded the facilities for operating a processor farm. One useful addition has been the provision of an operation **net_broadcast(length, message)** to enable the master to initialize all the workers, each of which uses the **net_receive** operation to read the message, packet by packet.

Another improvement has been to permit mixed configurations in which subsidiary networks (flood-filled or otherwise) are attached to the edges of the main network; thus, certain transputers on the periphery of the main network act as hosts to the subsidiary networks. The configuration file can also cater for processors other than transputers.

3.7 Retrospect and prospect

This chapter has concentrated on introducing the newcomer to the transputer, step-by-step: porting sequential programs, specifying the configuration, programming with multiple processes, writing schedulers, multiplexers and collators, and using processor farms. The aim has been to make the transition from the notions and mechanics of sequential programming to the notions and mechanics of parallel programming as painless as possible and, for this reason, the three most widely used languages have been adopted, together with the most widely used system. There is more to 3L's products than this chapter suggests, but the features described here are those that are safely usable; many of the other facilities should probably not be used at all. There is not yet anything approaching a standard in transputer software, except perhaps for occam. The communication and configuration features illustrated in this chapter have their counterparts in the offerings of other systems such as those of Meiko and Parsys and the Helios system; with these and others, care should be taken in deciding which features to adopt and which to avoid: the history of parallel programming is littered with irreproducible time-dependent bugs arising from the undisciplined use of unsafe synchronization and mutual exclusion methods. The consequence of this is a thriving market for software like 3L's Tbug to help you find out where your program has gone wrong.

Many programmers are loth to adopt truly parallel languages like occam and so there continues to be a demand for C, Fortran, Pascal and other languages that can be used to write sequential processes that can run concurrently with other processes. This gives a pattern of a few large processes. If one's application requires many small processes, occam should be used and many new problems and techniques for the efficient use of parallelism come into play: much of the literature on programming transputers is concerned with these. However, a great many applications can be implemented perfectly efficiently and easily using a few large processes – some of these are described in later chapters. The main consideration in guaranteeing the effectiveness of such systems is ensuring that the amount of computation considerably exceeds the amount of communication. Where fine grained parallelism is required, it may be easiest to use a mixture of occam processes for the critical parts that require considerable interprocess communication and, say, C processes for the parts that are primarily sequential.

Part II

HARDWARE FOR PARALLELISM

Computer designers have continuously striven for increased computer power to solve ever more complex problems. This can be partly achieved by making the processor execute more quickly, however cost and limitations due to the physics of the devices can make this route prohibitive. A more fruitful direction lay in the concurrency of elementary operations where the fetching of an instruction could be overlapped with the execution of a previously fetched instruction. This was further enhanced by the introduction of instruction pipelines which have been implemented in many modern processors. It became apparent that to gain dramatic increases of computing power, multiple processors had to be employed. Systems were built which used processors designed for traditional computers; these employed a variety of stratagems to achieve parallelism, but failed to disguise that the individual elements were not designed for parallel processing. The transputer, a single chip component, was specifically designed for parallel computation. This device uses message passing to accomplish interprocess communication. Many transputers can be used to create a parallel system at a comparatively modest cost, offering the advantage that as the complexity of the solution increases, additional transputers can be utilized to reduce the overall execution time. Advantage can also be gained from the multiplicity of similar processing elements as it becomes feasible to tolerate hardware faults within the system while suffering the penalty of reduced speed of execution.

The two chapters comprising Part II describe the hardware from device level through to system level components, giving the reader an appreciation of the large range of transputer based solutions possible. The components described can be readily used to implement various topologies by employing the transputer links, thereby enabling the computer system to take the form of the solution. The forms can be so diverse as pipelines and trees through to hypercubes and cube connected cycles. Part II provides the reader with a knowledge of the hardware which is available to the system builder who is going to develop an application exploiting the parallelism offered by the transputer.

Transputer hardware (Chapter 4)

This chapter gives an overview of low-level transputer hardware by describing the architecture of the transputer family of devices. This includes a description of the links that are the distinguishing feature of transputers, which enables a variety of topologies to be configured. The various transputer devices are presented, and the TRAM and motherboard standards introduced. Readers are encouraged to interface transputer systems to the real world by a detailed description of interfacing techniques.

The chapter concludes with practicals employing link adaptors which interface external devices to a transputer system, making it feasible to perform the usual lighting of LEDS and achieve feedback control of a motor. These practicals formed part of a course given by the authors, and have the advantage that the communication with hardware is via channels, thus adhering to the philosophy inherent in transputer systems.

Manufacturers' hardware equipment (Chapter 5)

Building on the chip level description of Chapter 4, this chapter itemizes many of the board and system level components which can be combined to achieve complete transputer based systems. A range of general processing TRAMs are presented, as are those designed for specific standard interfaces. Where appropriate, the interface standards are described briefly to assist the reader in the decisions which must be made when configuring a complete system. Thus transputer based systems can be enhanced through integration with other computer equipment. Components of image processing systems are introduced as these form an important application area for transputer systems. Finally, there is a brief overview of some large systems which evinced that supercomputer performance can be achieved from relatively simple components. However, the nature of the marketplace is such that these offerings form part of a rapidly moving target, so that the systems and even the manufacturers do not remain constant.

P. J. Sweeney

Chapter 4
Transputer Hardware

J. W. Flanigan and W. A. Coey

4.1 Introduction

This chapter provides a brief overview of low-level transputer hardware. The origins of multiprocessor systems and the software origins of the transputer are discussed, and some building blocks and interfacing components are introduced. For detailed descriptions and guides to programming, the data sheets and technical notes from Inmos should be consulted.

4.2 Background and concepts

4.2.1 Conventional uniprocessor architecture

By the late 1940s, the foundations for digital computer design had been formalized by John von Neumann. The architecture was deceptively simple, comprising a single processing unit connected to a single store of memory, and one instruction was executed at a time.

This basic design has dominated digital computing since that time, and most personal computers and workstations still use the basic techniques. However, developments in hardware and software since the 1960s spawned systems in which the processor time and resources were shared between different programs (IBM Series 360), and saw also the introduction of 'vector' computers (Cray-1) in which separate circuitry handled arithmetic calculations in parallel.

The era of the microprocessor which commenced in the 1970s led to the introduction of microcomputers and workstations and saw increased use of coprocessors, multiport memory and multiprocessor systems. Several basic design methods evolved, leading to standard multiprocessor architectures.

4.2.2 Multiprocessor architectures

There are two basic methods by which one processor can share data with another. Data can be communicated directly, or somehow it can be made accessible by more than one processor.

One technique in the latter classification is to have **common** or **shared memory**. Here, a block of multiport memory is accessed by the processors, and contention hardware ensures that only one of the processors can update the memory at any time. This technique is effective when there are few processors in the system. However, as more processors are added there is more contention for the memory, leading to a bottleneck which limits the overall speed of the system.

An alternative to shared memory is to have a **common bus** for all processors, but this leads to the same type of contention problems.

Enhancements to the basic principles can improve performance of the above architectures. For example, each processor might have its own local memory and use global memory only when absolutely necessary. However, a radically different approach might be to remove the need for common resources. This was the case with the transputer.

4.2.3 The transputer

As outlined in Chapter 1, a transputer is a single chip 16- or 32-bit microprocessor with its own memory and high-speed communication links. It was developed by the Bristol-based company Inmos Ltd. Originally set up in 1979 to bring Britain into the microelectronics industry, Inmos produced semiconductor memory chips, and the revenue was used to fund the development of the transputer. The idea was to combine memory and processor on a single chip, and for these chips to be directly connectible. The name transputer was derived from TRANSistor and comPUTER, since the component was to be a basic building block like a transistor but a complete computer on a chip. The design materialized by 1981 and the first transputer, the T414, became commercially available in 1985.

Software origins

Sequential software consists of program instructions executed in sequence, where execution of a single instruction comprises two steps – fetch instruction and required data, then execute the instruction. Pascal, Basic and Cobol are some of the many examples of sequential programming languages, and the conventional microprocessor is best suited for sequential use in a single processor environment.

Concurrent software when running on multiprocessor systems involves the execution of a number of programs or processes simultaneously. Consequently, more instructions can be executed within a given time. Modula, Ada, Concurrent Pascal and occam are examples of concurrent programming languages. These languages were developed on sequential machines and executed by sharing resources and processor time.

Occam, based on Hoare's Communicating Sequential Processes, has three primitive processes – assignment, input and output, as mentioned earlier. These can combine with constructors – SEQuence, PARallel or ALTernative – to form larger processes, which may be connected to form concurrent systems. The transputer was designed to represent the structure of occam. A transputer executing an occam program is a hardware representation of an occam process, and a network of transputers can therefore be described as an occam program. Occam processes communicate by means of channels, where one process will output a message onto a channel and another process inputs that message from that channel. Channels are implemented on the transputer by the high-speed links. Channel communication is synchronized, so if a channel is used for input in one process and output in another, communication takes place when both are ready.

A single transputer can execute concurrent processes by sharing processor execution time between processes, and by implementing channels using internal memory locations. The time sharing is managed by a microscheduler which schedules and deschedules processes. When a new concurrent process is found it is added onto the end of a scheduled list. Concurrent processes on the list may be either active or inactive. Active processes are either being executed or waiting to be so, and inactive processes are waiting to input or output or for a predefined time to elapse. An important feature is that inactive processes do not use processor time. Concurrent processes may be given high or low priority. There are a number of rules which govern these priorities, but in general, a low priority process will not be executed if a high priority process is on the scheduled list.

The transputer, then, unlike conventional microprocessors, was specifically designed as a building block for parallel systems with its roots firmly in proven parallel software techniques. While conventional languages such as C, Fortran and Pascal can benefit from the high performance of a single transputer, these languages have been extended to make effective use the transputer's parallellism and communications mechanisms.

4.3 The chipset

4.3.1 General transputer architecture

The transputer architecture (Figure 4.1) defines a family of programmable VLSI (Very Large Scale Integration) components. A typical member of the transputer family of devices is a single chip containing processor, memory and high-speed communication links which provide point-to-point connection between transputers. Each transputer product contains application specific circuitry adapting it to a particular use.

Packaging Transputers are packaged pin grid arrays or for surface mount in PLCC J-bend format. Extended temperature devices comply with MIL-STD-883C.

Services All transputers require a single 5 V power supply and operate from a standard 5 MHz input clock. The clock is multiplied internally, thus avoiding the problems of distributing a high-speed clock signal.

Buses The transputer uses conventional address and data buses for accessing external memory. On some devices the address and data buses are multiplexed, however, little if any external logic is required due to the design of the internal memory interface. The memory interface is user configurable and directly supports dynamic and static RAM, as well as ROM and EPROM. Several kilobytes of fast internal RAM can be used for core processes or cache purposes. Direct Memory Access (DMA) via the memory interface is catered for by *MemRequest* and *MemGranted* signals.

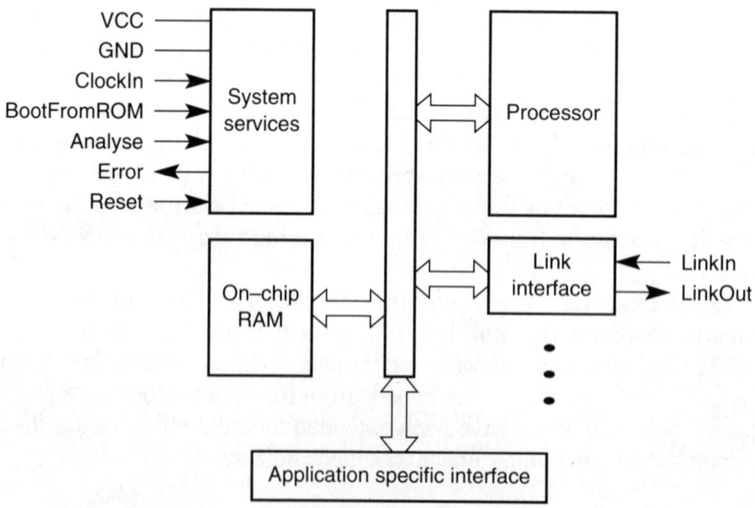

Figure 4.1 Transputer architecture.

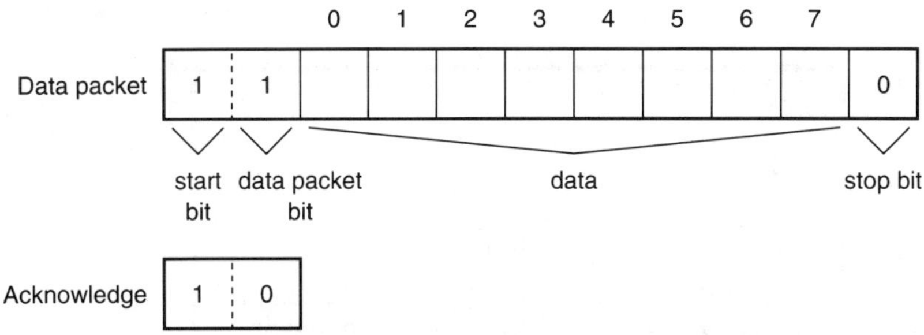

Figure 4.2 Link protocol.

Links Transputers can be arranged into network topologies where communication is by means of links which operate at 5, 10 or 20 Mbits per second. A link comprises two signals: *Linkin* and *Linkout*. The links are TTL compatible and have a nominal impedance of 100 ohms. Each byte of data is transmitted as a packet of 11 bits, made up of 2 start bits, 8 data bits and 1 stop bit. For each byte sent, the sender waits for an acknowledge from the receiver before continuing the transmission. The acknowledge is a 2-bit message which can be sent as soon as reception of the data byte starts, and since the receiver has a buffer for the next byte, the transmisson is continuous with no delays between data bytes (Figure 4.2). Data packets and acknowledge packets are multiplexed on the links. When there are relatively few transputers in a network it is feasible to manually interconnect the links, which will typically be brought out via short twisted pair cables. However, manual connection is both time consuming and error prone, and support chips to automate the process are available.

Subsystem Other signals such as the Reset, Analyse and Error lines are daisy-chained through multiprocessor systems for control and debugging purposes, where the debugging process involves halting transputers to examine the state of processes, pointers and data registers.

Booting At power on or after a reset, the transputer can be bootstrapped from either a link or external ROM, selection being made by the appropriately named 'BootFromROM' pin. Typically, a network will be loaded from a single host with code for each transputer sent in packet format through the network.

4.3.2 Some real transputers

A variety of transputers are available. This ranges from a low cost 16-bit device (T222, Figure 4.3) to a full 32-bit processor with on-chip floating point unit (for example T805, Figure 4.4). Figure 4.5 shows the internal architecture of the T805.

The figures are extracted from Inmos literature. Full engineering data is available from Inmos Ltd, who are acknowledged for their kind permission to reproduce the extracts.

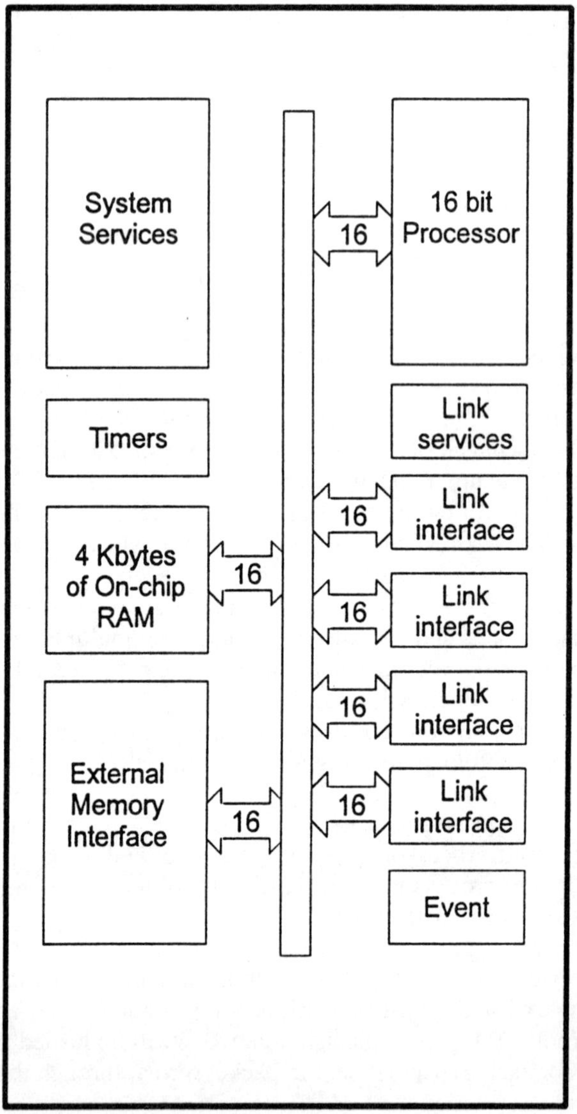

Figure 4.3 T222 transputer. *Source:* INMOS Limited engineering data. INMOS is a member of the SGS-THOMSON Microelectronics group.

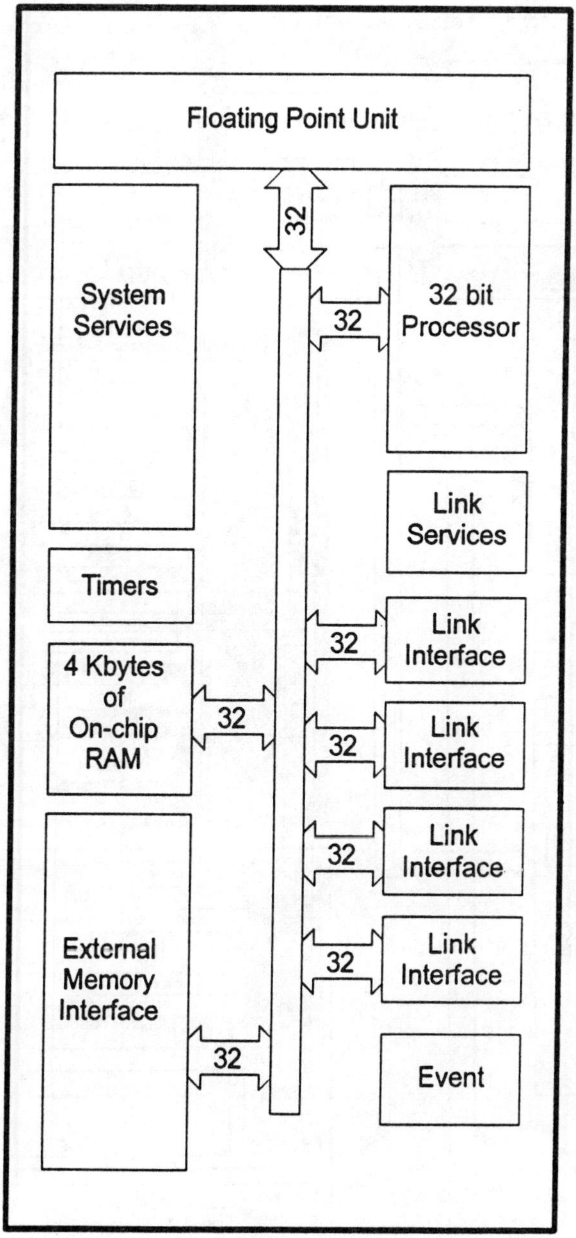

Figure 4.4 T805 transputer. *Source:* INMOS Limited engineering data. INMOS is a member of the SGS-THOMSON Microelectronics group.

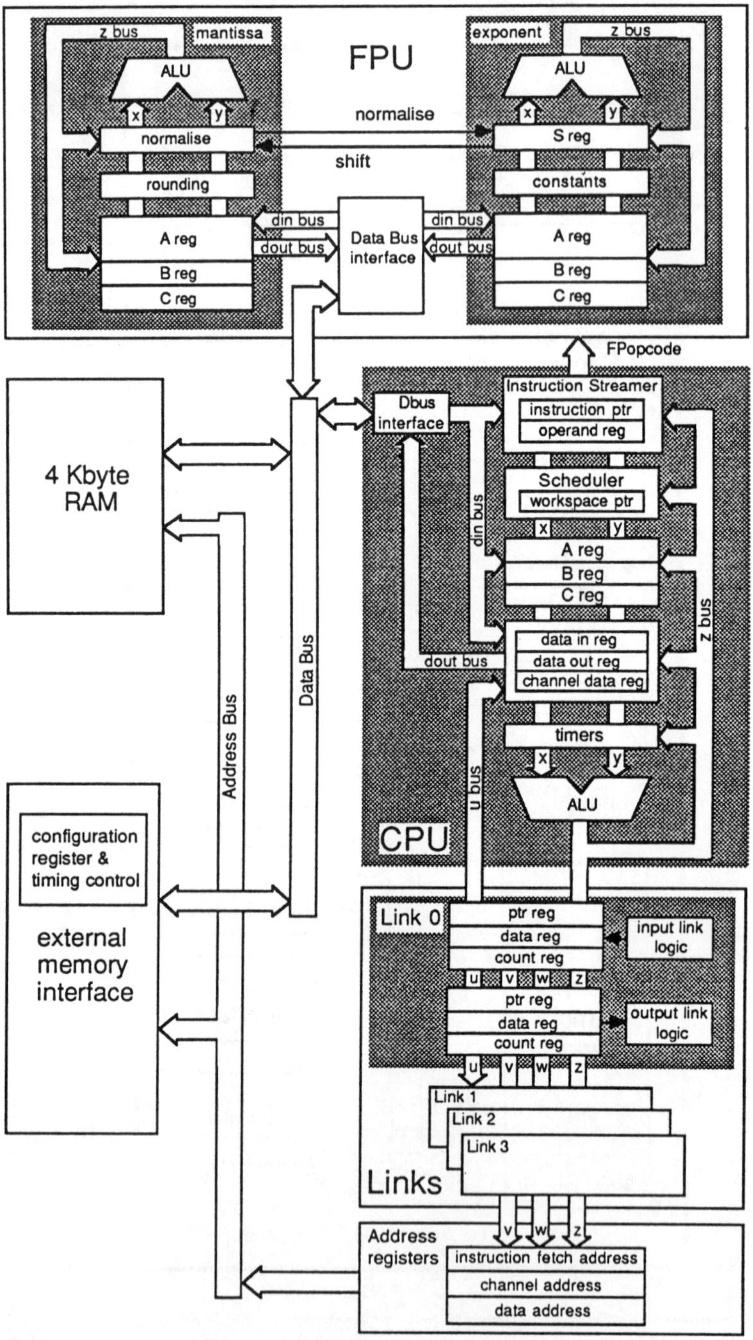

Figure 4.5 T805 internal data paths. *Source:* INMOS Limited engineering data. INMOS is a member of the SGS-THOMSON Microelectronics group.

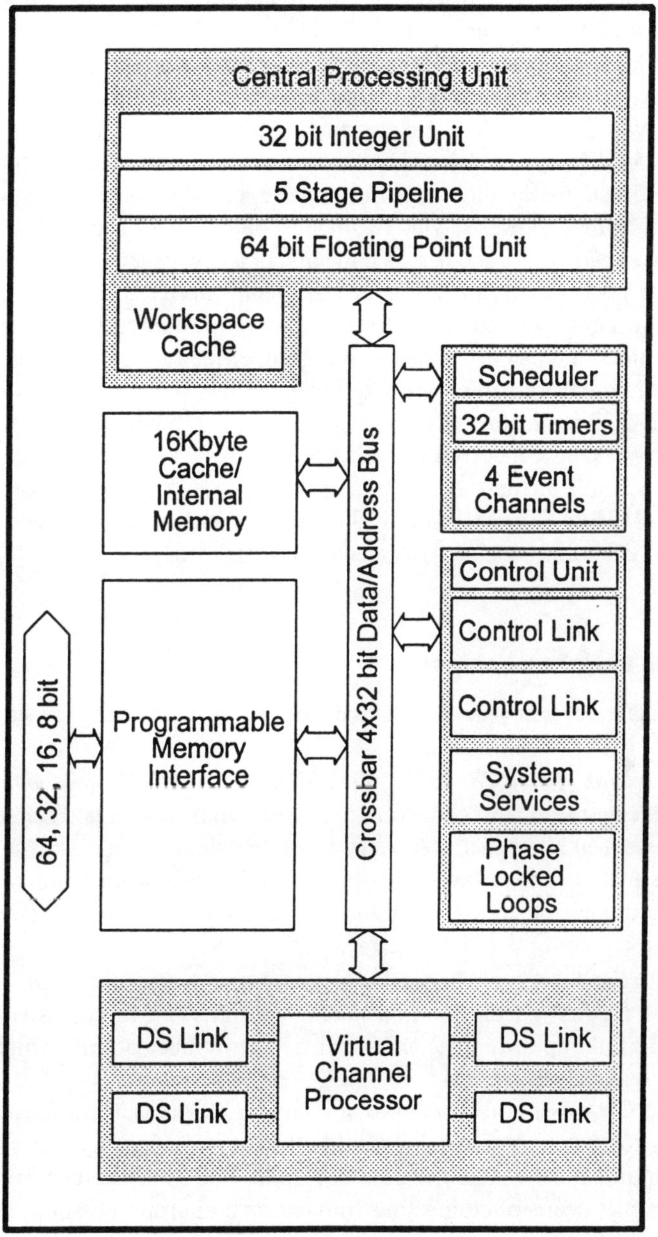

Figure 4.6 T9000 second generation transputer**.** *Source:* INMOS Limited engineering data. INMOS is a member of the SGS-THOMSON Microelectronics group.

Development of transputer technology has led to a second generation of devices with further improvements in performance of both the processor and the links. The follow-up to the T800 series transputer was a project to develop a second generation transputer, codenamed the H1 and later to be marketed as the T9000.

Under the ESPRIT programme, the PUMA (Parallel Universal Machine Architecture project to develop the H1 and a new communication link routing chip, the C104, was undertaken by a consortium led by Inmos. Partners in the consortium included Siemens and a number of European universities. The codename H1 allegedly arose while the consortium were meeting in a hotel where 'Yes Minister's' Humphrey was staying (Humphrey always got it right!). The T9000 was announced in early 1991 and heralded the first of a new generation of transputers.

The T9000 facilitates upgrades to much improved performance both in terms of processor speed and interprocessor communications. The improvements included the use of an instruction grouper feeding a five stage instruction pipeline in which multiple optimized instructions are executed in each clock cycle. The communications system was redesigned to become a virtual channel processor working concurrently with the main processor and supporting faster links. As a result of these developments, the processor peak performance rose from 30 MIPS to 200 MIPS, and the link speeds increased from 20 Mbits/second for each link to 100 MBits/second (Figure 4.6).

4.3.3 The peripheral chipset

Each generation of transputers required the development of a range of support chips.

First Generation (T2,T4,T8) C011 Link Adaptor The C011 link adaptor (Figure 4.7) provides a means of connecting transputer links to standard microprocessor and subsystem architectures. It converts the bidirectional serial link data into parallel data streams, and allows full duplex communication. Data reception is asynchronous and is therefore independent of clock phase between devices.

C004 Crossbar Switch The C004 (Figure 4.8) is a transparent programmable link switch for first generation transputers, and provides full crossbar switching between 32 link inputs and 32 link outputs. The number of link switches can be increased by cascading C004s.

The C004 is not restricted to computer configuration applications, since the ability to switch dynamically would allow it to be used as a message router in a communication exchange. The C011 and C004 may be combined to provide a flexible means of connecting conventional microprocessor systems.

In addition to transputers, Inmos manufacture a range of other semiconductor devices such as Colour Look Up Table (CLUT) chips and Digital Signal Processing hardware.

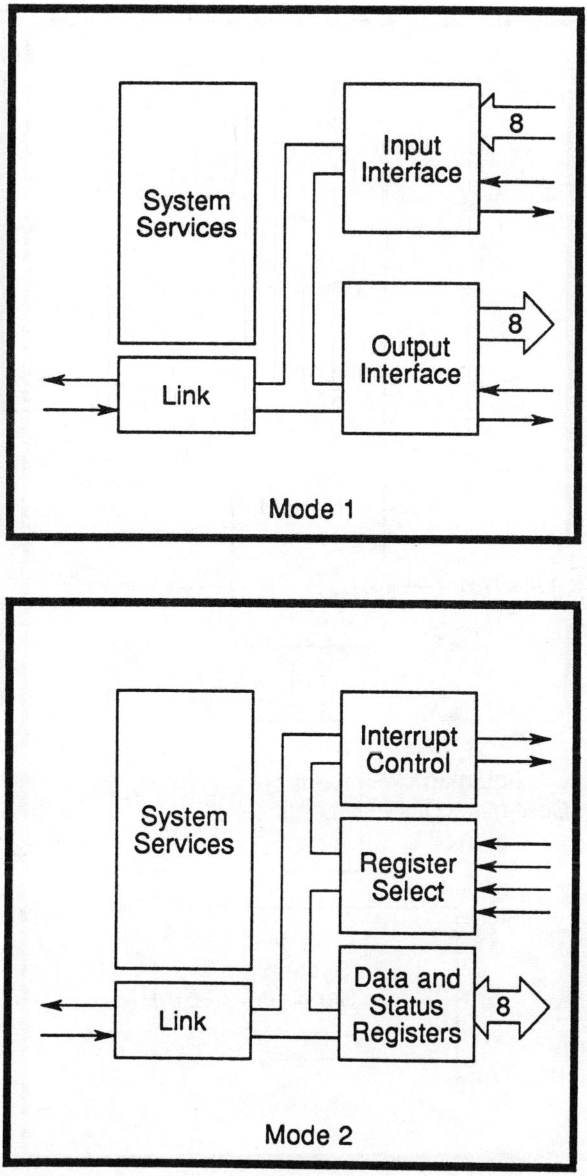

Figure 4.7 C011 link adaptor. *Source:* INMOS Limited engineering data. INMOS is a member of the SGS-THOMSON Microelectronics group.

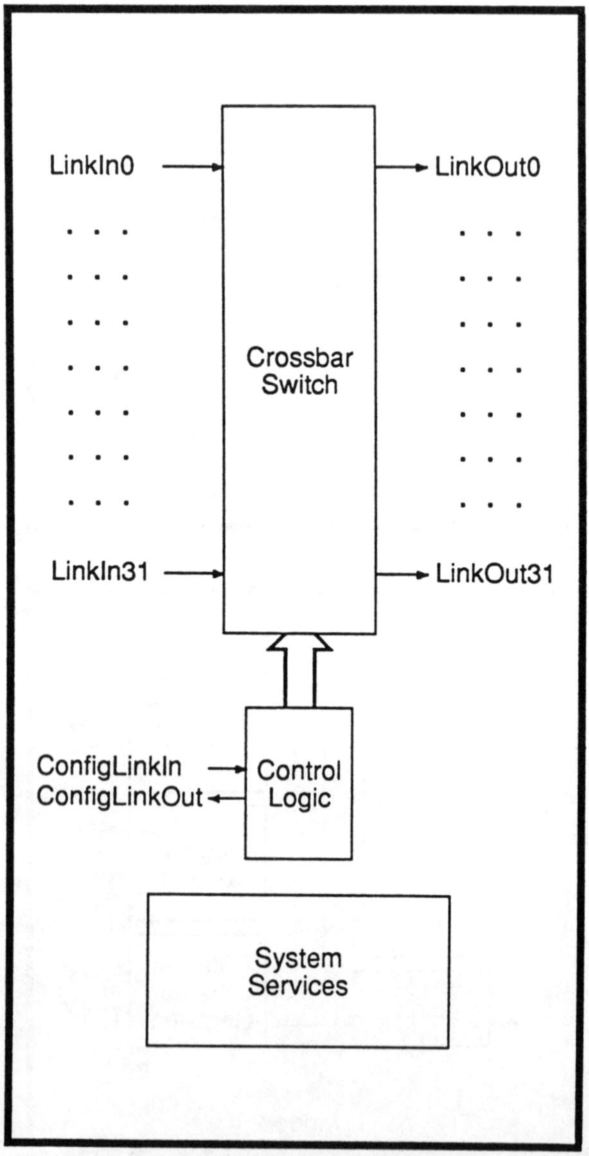

Figure 4.8 C004 programmable link switch. *Source:* INMOS Limited engineering data. INMOS is a member of the SGS-THOMSON Microelectronics group.

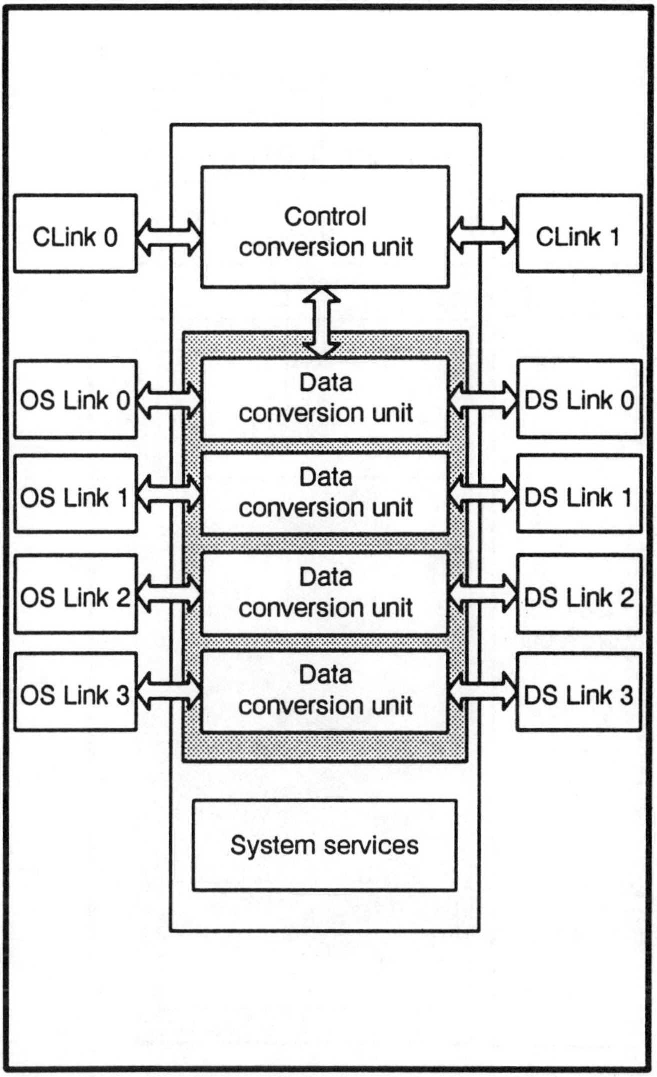

Figure 4.9 C100 protocol converter. *Source:* INMOS Limited engineering data. INMOS is a member of the SGS-THOMSON Microelectronics group.

Second Generation (T9) The higher performance links of the T9000 required the development of a range of products for direct communication with first generation devices. Two products were announced – a link protocol converter (Figure 4.9) and a routing chip (Figure 4.10). The protocol converter simply provided a means to connect the physical links and subsystem of a T9000 with those found on the T2/T4/T8 series transputers; however, the routing chip significantly enhances the use of transputer links in general.

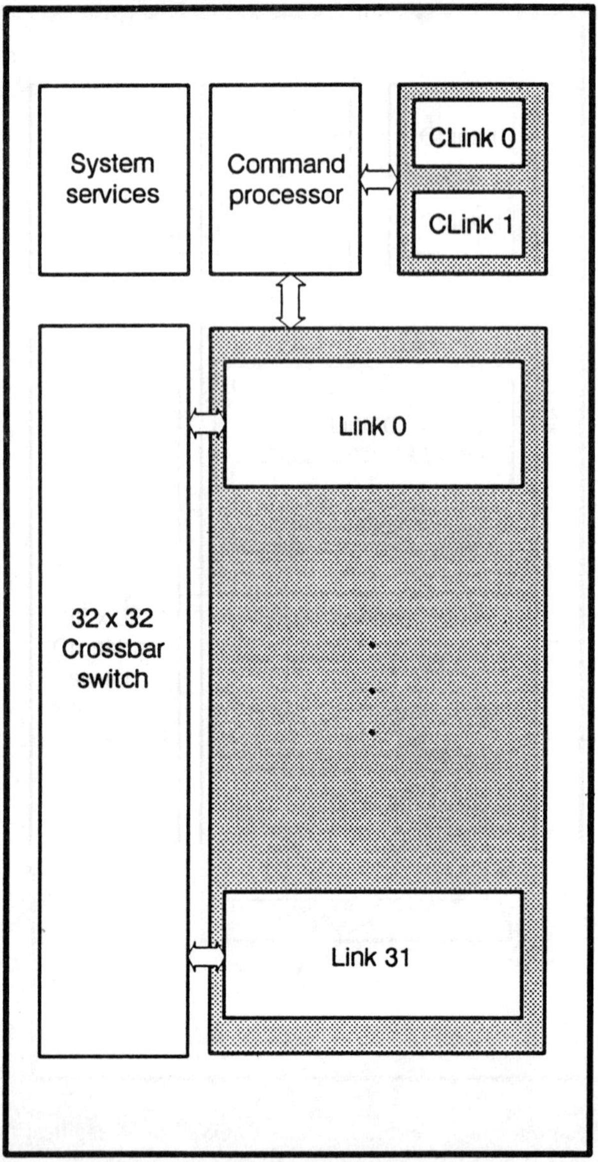

Figure 4.10 C104 packet routing switch. *Source:* INMOS Limited engineering data. INMOS is a member of the SGS-THOMSON Microelectronics group.

The C104 routing chip implements packet switching for transputer networks. The chip is a 32 × 32 way cascadable crossbar switch, and it enables the construction of networks in which direct connection between each of the devices is emulated.

4.4 The TRAM and motherboard standards

4.4.1 The need for a standard

The transputer was designed as a building block, and it was inevitable that modular sub-assemblies would evolve. One of the earliest building blocks was on the Inmos B003 quad transputer board (Vadher and Walker, 1987). The B003 was designed in its strange format with more than half the board devoid of components to emphasize the simplicity and economy of circuit design using transputers. The B003 contained four repetitions of a simple circuit comprising a T414 with 256 kbytes of RAM.

Most sub-assemblies would contain a transputer, a block of memory and links to other transputers, and it was important to standardize before piecemeal development led to an incompatible range of products. Fortunately, Inmos published a Module specification before this took place, and only one or two modules (based on the design of the B003) materialized before the standard was established.

4.4.2 The standards

Inmos adopted a mother/daughter board approach on which to standardize. The earliest commercial product was the B006 board (Hoare, 1978) which housed modules conforming to the Inmos 16-pin module specification. This board was a double eurocard designed to slot into the Inmos Transputer Evaluation Module (commonly known as the ITEM rack), and could accommodate eight modules in addition to the T212 on board. Since then, the Module Motherboard Architecture (Vadher and Walker, 1987) and Dual-in-line TRAM specifications (Inmos Databook series) have been published, and the standard adopted by almost all manfacturers of transputer-based kit.

The modular approach to hardware design makes good economic sense, and provided that the module design is well thought out, it becomes relatively easy to engineer a diverse product range for a wide number of applications. The Inmos architecture permitted the flexibility required both in mechanical and electrical terms. It is possible today to engineer a system using a mix-and-match approach using modules of different sizes and specifications.

The modular nature of transputing has enabled those who would not previously have been concerned with hardware issues to engineer their own computer systems. It has also simplified matters for those engaged in hardware design. For example, the major concern in the development of a multiprocessor accelerator board for a PC is in the design of the interface to the bus, and not with the complexities of the accelerator itself.

The apparent simplicity exists largely because Inmos got it right.

4.4.3 Dual-in-line TRAMs

TRAMs (TRAnsputer Modules) are sub-assemblies of transputers (or other components with Inmos links). TRAMs have a standard dual-in-line pinout and are manufactured in a range of specifications and sizes. The earliest TRAMs comprised a transputer and 32 kbytes of static RAM. However, progress in semiconductor and surface mount technologies has enabled the smallest module to contain a transputer and several megabytes of RAM.

The basic unit is a size 1 TRAM, which is about half the size of a credit card. The other sizes are 2, 4 and 8 (the largest module specified). A size 8 TRAM conveniently fits onto standard PC and VME boards.

The secret of the success of the TRAM lies in its simple standardized interface. The basic building block has its functionality on-board, and the interface can thus be limited to 16-pins (Figure 4.11).

The interface is such that the TRAM standard is independent of module functionality, transputer type, word length, speed and packaging.

- *Link*.* Pins* are the Inmos links which may be buffered on the module.
- *VCC and GND* provide the standard 5 V regulated supply to the module circuitry.
- *Clockin* provides the standard 5 MHz clock input to the transputer on the module.
- *Reset* and *Analyse* are inputs connected from the TRAM pins directly to the corresponding pins on the transputer contained on the module.
- *Error* is an output originated from a transputer on the module. This signal is wire OR'ed with other Error outputs, and is usually driven by open collector circuitry on the TRAM.

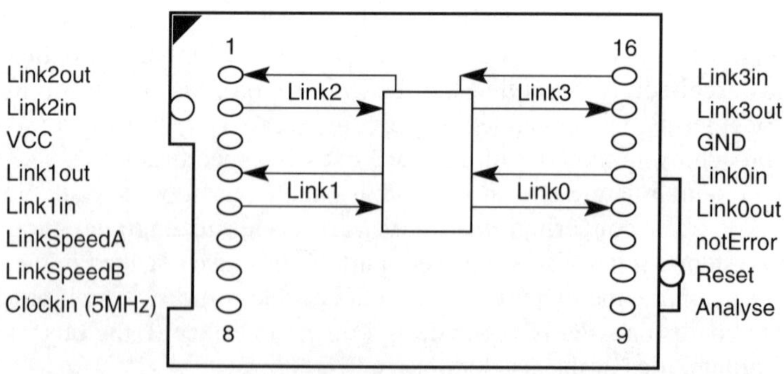

Figure 4.11 TRAM pinout.

Subsystem

To cater for hierarchical control of modules, an additional set of connections *may be present* on a TRAM. This **subsystem port**, as it is called, is incorporated on the underside of a TRAM, and consists of three signals – *SUBSYSTEM RESET, SUBSYSTEM ANALYSE* and *SUBSYSTEM ERROR.*

The motherboard will have a corresponding set of connectors, and a standard pin header is required to connect the two, if necessary.

Any additional application specific connections such as video or disk signals are made directly to the module. The recommended means is by a strip of pins on a 0.1" grid to accommodate standard strip cable connectors.

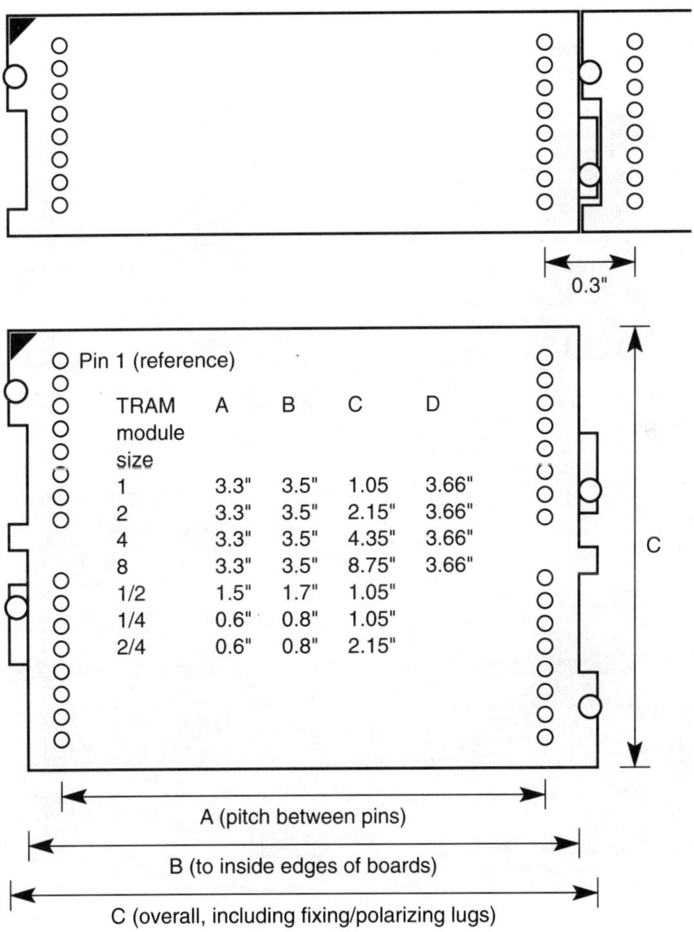

TRAM module size	A	B	C	D
1	3.3"	3.5"	1.05	3.66"
2	3.3"	3.5"	2.15"	3.66"
4	3.3"	3.5"	4.35"	3.66"
8	3.3"	3.5"	8.75"	3.66"
1/2	1.5"	1.7"	1.05"	
1/4	0.6"	0.8"	1.05"	
2/4	0.6"	0.8"	2.15"	

Figure 4.12 TRAM physical dimensions. *Source:* INMOS Limited engineering data. INMOS is a member of the SGS-THOMSON Microelectronics group.

TRAMs can be stacked. Where a TRAM occupies more than one site for mechanical reasons (for example, to accommodate additional circuitry such as memory on the TRAM), the standard allows for modules to be stacked. The mechanical needs for TRAMs do not therefore interfere with electrical functionality. Figures 4.12 and 4.13 show some of the mechanical characteristics for TRAMs.

Mechanical retention of TRAMs, while not usually required in the office PC, may become necessary where ruggedness is required, for instance in an industrial environment, or when other connections are made directly to the TRAM.

Most TRAM circuit boards have now been engineered to include fixing holes to accommodate nylon bolts with which to secure the modules. Figure 4.14 shows a size 8 TRAM with subsystem connector and fixing holes.

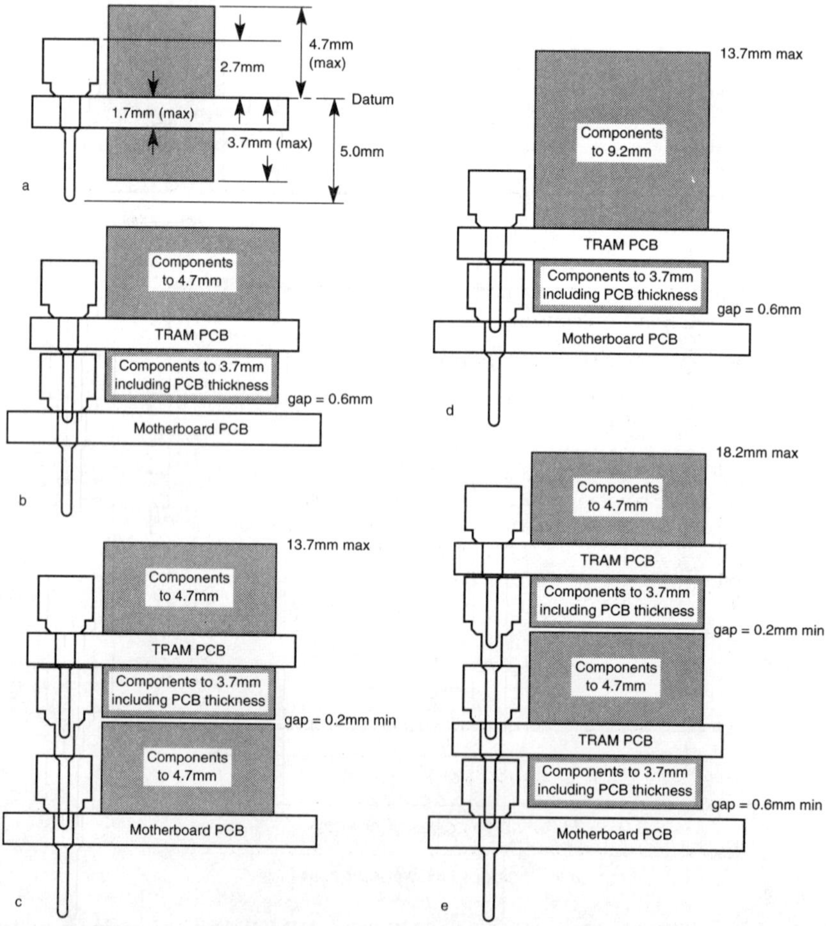

Figure 4.13 Stacking TRAMs. *Source:* INMOS Limited engineering data. INMOS is a member of the SGS-THOMSON Microelectronics group.

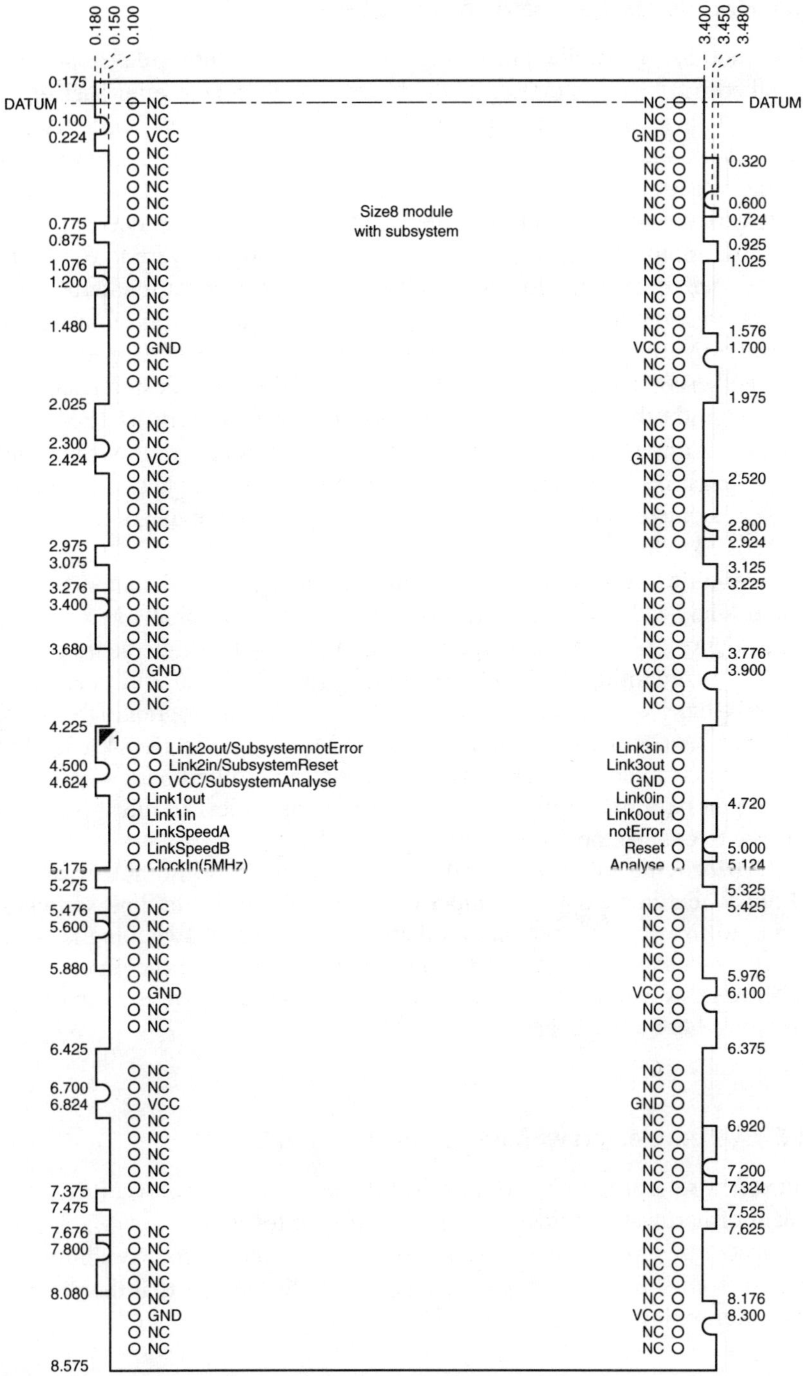

Figure 4.14 Size 8 TRAM with subsystem. *Source:* INMOS Limited engineering data. INMOS is a member of the SGS-THOMSON Microelectronics group.

4.4.4 Module motherboard architecture

To use modules as parallel processing building blocks, Inmos developed a range of motherboards. The simplest motherboard would provide a number of identical, independent sites for modules and the necessary common circuitry such as power, clock and subsystem signals. However, to provide a system which can link with a host computer and where the links between the various modules in the network are software configurable, additional features are required.

The Inmos design goals for module motherboards were to ensure diversity of system design and to maintain the modularity even at motherboard level. The aim was to devise a system in which any number of transputers could be combined in any configuration, and the resulting products (for example, the B008 motherboard (Inmos Databook series)) fulfilled these design goals.

A motherboard like the B008 has an interface with its IBM PC host computer. Although the detail of the interface is not covered by the standard, the host must be able to communicate with the transputer network, and this is generally achieved either using shared memory or by the use of a link adaptor as described in Section 4.5.3.

Modules are connected in a pipeline using links 1 and 2 from each module. The head of the pipeline (the pipehead) and the pipetail are brought to an edge connector so that the pipe may extend beyond a single board.

The remaining two links (0 and 3) from each module are taken to IMSC004 link switches on board. The link switch is a programmable crossbar switch for connecting links, and is described briefly in Section 4.3.3. The link switch is controlled by another transputer on the motherboard which is itself part of a separate 'control' pipeline. Inmos provide Module Motherboard Software to control the link switches.

To provide greater flexibility when connecting motherboards, some additional features exist. A number of spare links from the link switches are brought to the edge connector, and therefore multi-level switching is possible. Hierarchical systems can be built using the first module of the pipeline on any motherboard to control a subsystem of other transputers on the same motherboard, or other motherboards.

4.4.5 System design with motherboards and TRAMs

Engineering a computer system or embedded system was traditionally the domain of hardware engineers, and the software for the system which evolved was written by a programmer. This was the case even when a system was constructed from subsystem modules, usually due to the fact that there remained substantial wiring, interfacing and other hardware issues.

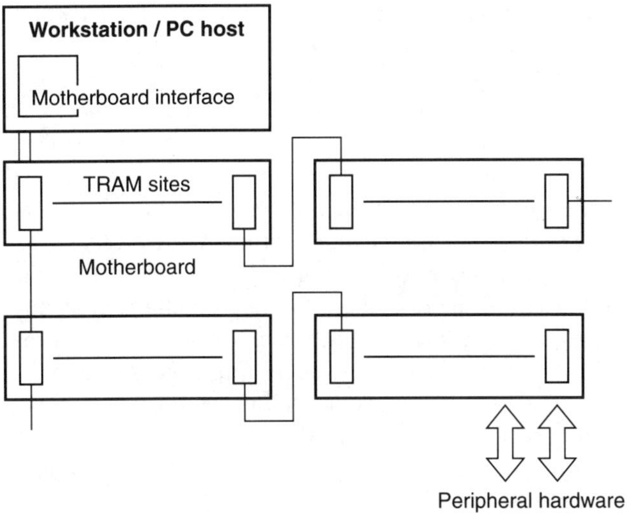

Figure 4.15 Subsystem development using motherboard and TRAMs.

However, transputing and its building blocks, from silicon to system level, has shifted the boundaries somewhat. With the general trend towards system simulation even for relatively small engineering projects, transputer-based development systems were ideal vehicles. A transputer could simulate a transputer network, parallel processes could simulate events in external hardware and, using results collected from the simulation, it was possible to predict where more or less processing power was required. It soon became the case that programmers dictated the hardware design and, due to the simplicity of the building blocks, the hardware engineering input was reduced to interfacing and embedding the system if the need arose.

System design using motherboards and TRAMs has therefore become a role of the software engineer. In keeping with the modular nature of transputing, a system is likely to be developed using subsystem components as in Figure 4.15. TRAMs for many functions have been manufactured by Inmos and others. Their scope includes general purpose Input/Output, Digital to Analogue and A to D conversion and interfacing to other standard buses and hardware. It is therefore feasible for the software engineer to construct an entire system around a PC- or workstation-based development system with a motherboard and appropriate 'off-the-shelf' TRAMs.

4.4.6 Limitations of the standard

There are few limitations imposed by the Module Motherboard and Dual-in-line TRAM standards. It has been found that not all possible network configurations can be realized when using a fixed pipeline backbone, as on the B008 motherboard. However, the problem can be overcome by using additional off-board switching.

The most common failures in hardware are interconnections. VLSI has greatly improved reliability in consumer electronics, partly due to reduced heat dissipation, but largely due to reduced component count and the resultant fewer number of interconnections. While the motherboard and TRAM standards recommend high quality gold plated connectors, the system is based on the interconnection of modules. For development work this is fine, but it will not lend itself readily to harsh industrial environments and the embedded system is unlikely to be constructed using the motherboard and TRAM approach.

The TRAM standard has been successful largely due to its simplicity. However, the fact that only the transputer's links and reset circuitry are brought to the edge of the module prevents any other method of connection to the transputer. Direct connection to the transputer's buses is thus possible only on a suitable TRAM or custom-built board.

4.5 Interfacing transputers

Interfacing transputer hardware with the 'real' world is probably a less daunting task than can be the case using other microprocessors. It may be possible for the software engineer with little hardware experience to develop an interface using readily available TRAMs, and for those who prefer to wield a soldering iron the job can be simplified by using a prototyping TRAM and traditional development techniques.

4.5.1 Interfacing using TRAMs

For developmental or experimental purposes, the easiest route by far to interface transputers with other hardware is by using TRAMs. TRAMs for a wide range of functions are now available from Inmos and other manufacturers. Analogue to Digital and Digital to Analogue converters are available, as are TRAMs to interface with standard bus architectures such as Small Computer Systems Interface (SCSI). These TRAMs are usually provided with a small library of routines, and perhaps with some demonstration software which in many cases can be adapted for the task in hand.

Connection to the external devices from a TRAM can often become the tatty bit of the system when interfacing using TRAMs. In an experimental system it is usually quite acceptable to have covers removed from machines for a hardware 'lash-up', but this is far from satisfactory outside the laboratory. An external rack-mounted motherboard is possibly the best option when interfacing using TRAMs.

4.5.2 Prototyping and test equipment

Conventional prototyping methods such as wire-wrapping can be used with transputer circuitry since the high clock rates are inside the transputer. Links with speeds of 20 Mbits/second need not cause problems if care is taken with routing and cable length. Use of prototyping boards with ground plane is essential.

On a wire-wrapped development the pin grid array transputer packaging can mean that the greatest density of wire will be around the 68- or 84-pin transputer. Inmos supply a prototyping TRAM with this part of the circuit completed, and its use could simplify an interfacing task and improve reliability of the module.

It is possible to lay out a double-sided printed circuit board containing a single transputer and some other components; however, it rapidly becomes evident why most commercial boards, especially those with memory, are multi-layered. Inmos recommend that a layer be used solely for power distribution on the board.

Traditional test equipment such as oscilloscopes and logic analysers will inevitably be required during serious development work; however, if the rules on the datasheets are observed (for example, clock rise times) a simple logic probe can be the only test equipment necessary for simple interfacing tasks.

With other microprocessors, emulators were an integral part of any development system; however, with transputers emanating from software rather than hardware roots it was perhaps not surprising that these were not seen as a priority.

4.5.3 Chip-level interfacing

Some application specific transputers were produced as part of the first generation. An example of this is the M212 disk controller transputer, which provided a standard Shughart interface. However, general purpose interfacing can be handled in two ways as discussed below.

Memory mapping The most common method of interfacing external hardware to a conventional microprocessor is by memory mapping components or registers into the address space of the microprocessor. This is perfectly feasible with transputers.

Some transputers (for instance, the T222) have separate Address and Data buses, while others have multiplexed buses. Those with multiplexed buses have a programmable memory interface incorporated in their design. Although primarily intended to simplify connection of dynamic RAM to the transputer, the signals can be used for a variety of memory mapping tasks. The interface has preset and programmable configurations to ensure correct timing of data and relevant control signals.

A preset configuration is chosen from the set by wiring an appropriate address line to the *MemConfig* pin of the transputer, or a user configuration can be supplied from a ROM or PAL.

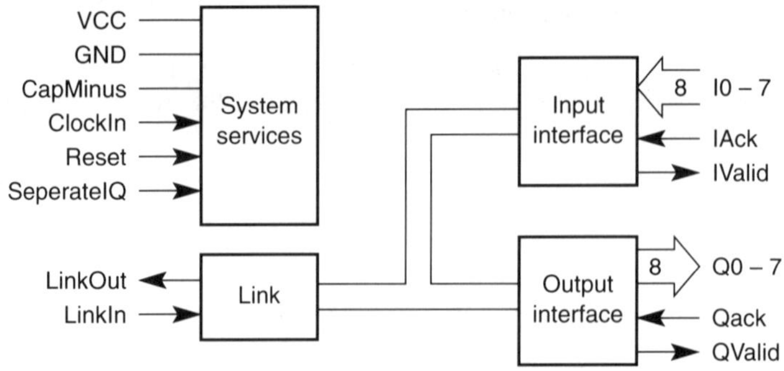

Figure 4.16 IMS C011 Link Adaptor – Mode 1. *Source:* INMOS Limited engineering data. INMOS is a member of the SGS-THOMSON Microelectronics group.

Link adaptors Link adaptors such as the C011 provide a convenient means of interfacing with transputers. The connection with the transputer network is via a standard Inmos link at 5, 10 or 20 Mbits/second corresponding to two channels, and the interface provided is a parallel data stream with handshake signals.

Since the interface to the transputer network is via a standard link, this method of interfacing facilitates simulation of the external hardware by another transputer (or process within one transputer).

The C011 can operate in two modes: Mode 1 – Parallel Input/Output; and Mode 2 – Microprocessor System Bus. In Mode 1, the link adaptor converts between a link and two 8-bit interfaces, one input and one output (Figure 4.16). The 8-bit interfaces each have two independent handshake lines to control data flow. The C011 used in this mode can provide a convenient means of connecting simple Input/Output devices or 8-bit Analogue to Digital converters.

When the link adaptor is not busy the handshake lines IVALID, IACK, QVALID and QACK are low. On data input, the external device places the data at pins I0–I7 and sets IVALID high. The link adaptor sends the received data out through the serial link, and when this is acknowledged the IACK handshake signal is taken high. The external device must then take IVALID low, at which point the link adaptor returns IACK to low completing the handshake for a data transfer operation. Similar handshaking takes place for output from a link. Data from the link is presented on output pins Q0–Q7 and QVALID is set high. After reading the data the external device sets QACK high. The link adaptor then sends an acknowledge down the link and returns QVALID to low. The external device must then set QACK low.

Two link adaptors in this mode and connected back to back at the parallel ports may be used as a frequency changer between different speed links.

In Mode 2 the link adaptor acts as an interface between a link and a microprocessor style bus (Figure 4.17). This is generally the means by which a

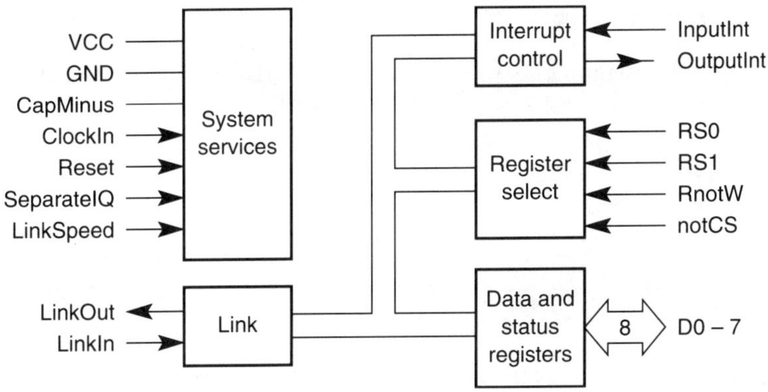

Figure 4.17 IMS C011 Link Adaptor – Mode 2. *Source:* INMOS Limited engineering data. INMOS is a member of the SGS-THOMSON Microelectronics group.

transputer network is interfaced to a host computer. Typically, the C011 registers will be memory mapped into the address space of the host.

Data is transferred to the microprocessor on lines D0–D7. Data transfer is via four registers – Input Status, Output Status, Read Data and Write Data. These are selected using lines *RS0*, *RS1* and *RnotW*, and the chip is enabled by *notCS*.

The Input Status register contains two flags, Data Present and Interrupt Enable. When data is received on the link it is placed into the Read Data register and the Data Present flag of the Input Status register is set high. This is combined with the Interrupt Enable flag, if high, to set the InputInt line high, which may be used to interrupt the microprocessor to activate a read data routine. Alternatively, the microprocessor can poll the status register testing the Data Present flag to ascertain whether data has been sent from the link. When the microprocessor has read the contents of the Read Data register the Data Present flag is set low.

The Output Status register contains two flags: Output Ready and Interrupt Enable. Data in the Output Data register is transferred down the link. When the adaptor is sending Data from the Output Data register down the link, the Output Ready flag is taken low and will remain so until an acknowledge is received from the link. If the Interrupt Enable flag is set high and the Output Ready flag is high, the OutputInt line will be set high and may be used to interrupt the microprocessor to activate a write data routine. Alternatively, the Output Status register can be polled to determine when to write data to the Ouput Data register.

The C011, then, is a versatile chip which provides a convenient means of interfacing external hardware with transputer systems. These can be cascaded to accommodate data transfer which is wider than eight bits. However, some additional logic will be required to combine the separate handshake signals.

4.6 Assignments

The following assignments formed part of a Hardware Technical Details course in the Northern Ireland Regional Transputer Support Centre, and provided a useful exercise in parallel programming and the use of link adaptors for simple input and output. Programming was permitted in Pascal or C although the use of occam was encouraged.

4.6.1 Transputer hardware practicals

These practicals use a multi-function experimental hardware board (Flight Electronics Ltd) which has been interfaced using two link adaptors to a transputer system installed in the IBM PC (Figure 4.18). A suite of development software on the PC provides an environment in which programs destined to run on a transputer (or network of transputers) can be developed in C, Fortran, Pascal and occam.

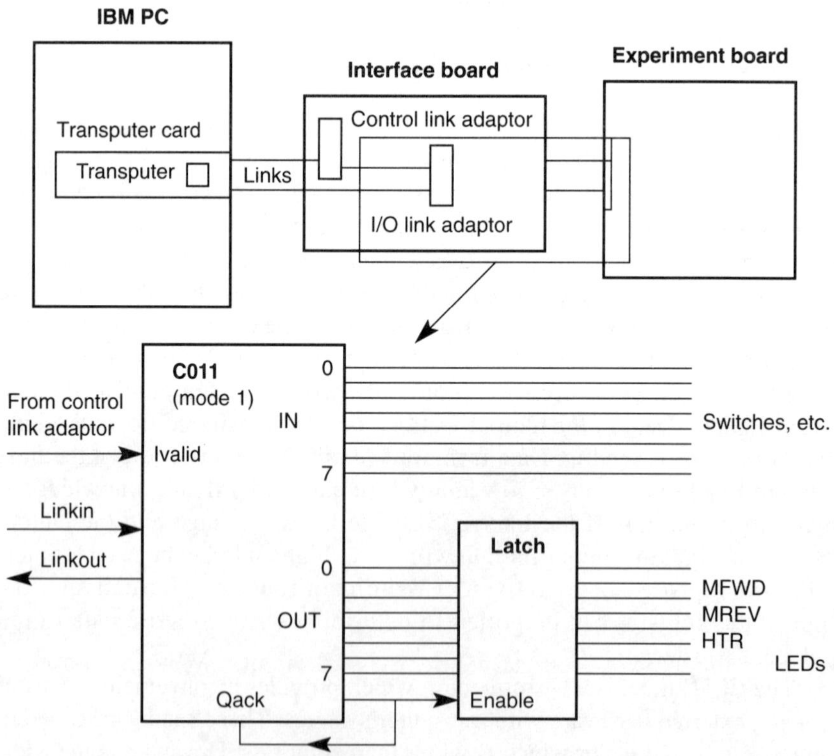

Figure 4.18 The experimental hardware.

The experimental board comprises a number of interconnected elements which can be used to model a variety of open and closed loop systems. As Figure 4.18 shows, a link adaptor which is basically a serial–parallel/parallel–serial converter has been used as the Input/Output interface component. Link adaptors support the standard link protocol, and thus all communication with the experimental hardware is via two channels (that is, Linkout and Linkin).

Data sent to the link adaptor from the transputer will appear on the 8-bit parallel output port. Data present on the input port can be latched by strobing an Input Valid bit on the link adaptor. In a real application, the data might be strobed in by an 'end of conversion' signal from an ADC, but for experimental purposes, one of the output bits of a second link adaptor will be used to generate the Ivalid signal.

Note that while the use of a link adapter is not the only means of interfacing to the transputer, it has the advantage that communication with the peripheral hardware is via channels. It is therefore possible to simulate hardware in a 'process', perhaps while the real hardware is being constructed. (Even the potential failures in the hardware can be modelled within the process.) When the hardware is ready, the channel name can be associated with a real link without modification to the program.

Experiment 1: Simple Input/Output

Refer to Figure 4.19, a programming model for simple I/O on the experimental board. Note that when a byte is output to the link adaptor, it will cause the LEDs to display the binary pattern of the byte.

Output

Write a program which will repeatedly scan the keyboard and echo to the link the ASCII code for any character typed.

Confirm that the pattern displayed on the LEDs has the correct ASCII code.

Input

The switches on the experimental board are connected to the input port of the I/O link adaptor.

Refer to the timing diagram in the C011 data sheet. Note that the data will be available on the link when the Ivalid pin is taken high. It should be returned low after the byte has been read by the transputer. Therefore, to read the current switch setting, it will be necessary to write a '1' to the control link adaptor (to generate the Input Valid signal). After the switch value has been read, a '0' should be sent to the control link adaptor.

Modify the previous program to continually monitor switches 1–7 and light the corresponding LEDs. Arrange for the program to terminate when the switches are all on.

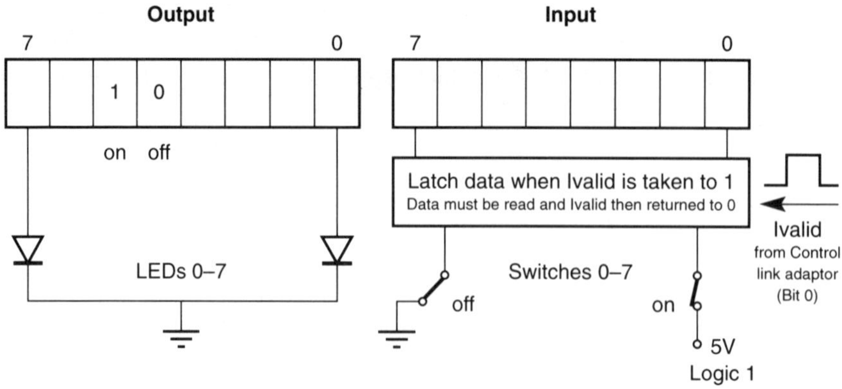

Figure 4.19 Programming model for simple I/O.

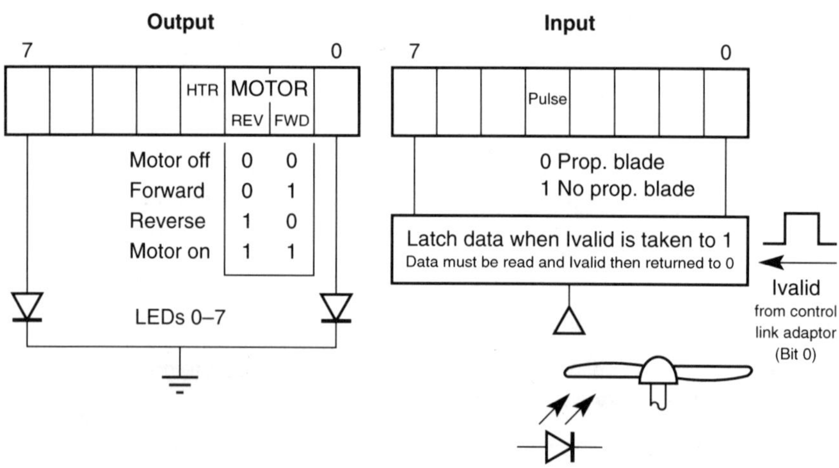

Figure 4.20 Programming model for motor control.

Experiment 2: Control

Refer to the programming model for Motor Control in Figure 4.20.

Note that output bits 1 and 2 control the motor on the experimental board, while input bit 4 indicates whether a blade of the propeller is cutting a light beam. The motor is a standard DC type, but may be proportionally controlled by altering on/off time.

Write a program which will control the motor and report its current speed in RPM.

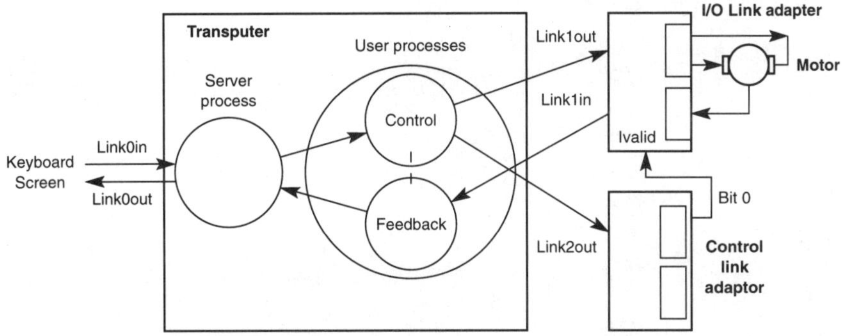

Figure 4.21 Program structure for Experiment 2.

Start the motor at approximately half speed (on for 50 msecs then off for 50 msecs).

The + and – keys should adjust the speed.

The R key should reverse the direction (after stopping the motor and delaying for one second).

The spacebar should cause the motor to stop and the program to terminate.

Figure 4.21 shows a suggested structure for the program.

Experiment 3: Closed loop control

Using the hardware in the same configuration as for Experiment 2, extend the last program or write a new one to include a new channel which closes the feedback loop (the dotted line in Figure 4.21). Arrange for the motor speed to remain constant (or close to it) whenever the propeller is slightly impeded.

References

Hoare, C. A. R. (1978) Communicating Sequential Processes. *Comm. ACM*, **21**(8), 666–677.

INMOS Databook Series. (1991) B008 IBM PC module motherboard product overview. In: *A Transputer Development and iq Systems Databook*, pp. 97–101. Inmos Ltd.

Vadher, A and Walker, P. (1987) IMS B003. *Design of a multi-transputer board.* Inmos Technical Note 10.72 TCH 010 00.

Chapter 5

Manufacturers' Hardware Equipment

P.J. Sweeney

5.1 Introduction

This chapter describes many of the available board and system level components which can be integrated to make systems of varying sizes. The components of such systems will be presented so that a system designer can envisage the feasibility of an intended solution. Emphasis will be laid on the components in three areas: general processing elements, TRAM support of standard interfaces and image processing components. Boards which enable the generation of a solution to a complete application will also be introduced. Inmos are, of course, the major manufacturer of TRAMs, but whenever another supplier is available this will be indicated. A list of hardware suppliers is furnished at the end of the chapter to enable the would-be system builder to obtain the latest product information. It is inevitable that any such description be fated to obsolescence, and this is particularly so when it is based on hardware components; nevertheless, it will enable a system designer to see the large range of possibilities available.

5.2 Background

Transputers can be found in a diverse range of computer products from embedded controllers to supercomputers. The ability to make powerful, scalable systems from a set of small modules is an attractive feature which has given the transputer an edge over its competition. The capability to use available computers as front-ends to transputer networks has enhanced their appeal, particularly on grounds of cost.

The development of transputers and their integration into existing computer systems highlighted the need for a standard interface to allow this activity to progress successfully. If a standard interface was not chosen it would have been mandatory on the designers to provide all of the desired configurations for each computer uniquely. The approach taken to minimize this design task, and stock holding, was to define a standard interface so that host specific boards could be produced. These boards could then easily support the transputer elements. The interface was chosen so that it was as simple as possible, yet supported the parallel computing envisaged; this interface was considerably helped by the fact that interprocessor communication is supported by four links. Therefore, the interface needed to have only 16 connections, eight for links, two for power and six for timing and diagnostics (Figure 4.11). This follows the traditional concept of specifying a simple interface and maintaining the complexity within the system components, and has enabled Inmos and other suppliers to have a range of host specific boards and a limited range of individual TRAMS.

5.3 General processing TRAMs

5.3.1 The TRAM standard

The TRAM standard (Inmos Ltd, 1991) has already been introduced in Chapter 4, which describes the pin-out and physical dimensions. It is therefore possible for designers to mix and match different sizes of TRAMs to suit their application, giving rise to a vast range of combinations. It was indeed convenient that the size of individual TRAMs, with size 1 being the smallest, was such that motherboards could be designed which allowed them to be readily integrated into standard desktop computers. A motherboard for an IBM PC can support up to eight size 1 TRAMs. It thus becomes possible to design systems which match the structure of the solution under investigation. The TRAM standard is independent of the transputer type, size of memory and function due to the interface being specified only at the link level. It is also possible to stack TRAMs above one another, but a combination of mechanical and thermal restrictions for some undertakings has led Inmos to lay less stress on this aspect of the design.

5.3.2 IMS B416 TRAM

The B416 TRAM is based on the IMS T222 transputer; this is an early version of the transputer, and is a 16-bit device with 16 address and 16 data lines. Thus the address space is 64 K in total, and is fully populated on this TRAM. The T222 has 4 Kbytes of internal RAM which occupies the first 4 K of address space, whilst the next 60 Kbytes are occupied by the external static RAM on the B416. The T222 runs at 20 MHz, which can achieve an instruction peak of 20 MIPS with a sustained throughput of 10 MIPS. The pin designations for the B416 TRAM follows the standard for a size 1 TRAM. The supply voltage should be in the range 4.75–5.25 V with the power consumption being 2 W. This device takes the form of a size 1 TRAM, with a component height above the printed circuit board (pcb) of 4.7 mm, with that below the pcb being 3.7 mm.

Since the B416 is based on the 16-bit version of the transputer, which has no inherent floating point capability, the uses for this TRAM are somewhat limited; however, it provides a low cost entry level into the world of parallel computing. The most appropriate applications are the routing or filtering of messages and the intelligent control of IMS C004 link switches.

Ordering details

Company	*Order No.*	*Description*
Inmos	IMS B416-10	64 KBytes SRAM – IMS T222-20
Inmos	IMS B402-10	8 KBytes SRAM – IMS T222-20
Transtech	TTM22	64 KBytes SRAM – IMS T222-20
Mark Ware Associates	TM264K1-1	64 KBytes SRAM – IMS T222-20

5.3.3 IMS B401 TRAM

The B401 TRAM can support the T414, T425 or T800 transputers which are all pin compatible. The T414 has been superseded by the T425, which has a faster internal cycle time (33 ns vs. 50 ns) and larger on-chip RAM (4 KBytes vs. 2 KBytes). The T425 and T800 each can provide a peak throughput of 30 MIPS, with the T800 offering on-chip floating point calculations. These processors have a 32-bit architecture with multiplexed address and data lines giving a potential 4 GBytes of address space (Figure 5.1). The T425 and T800 have 4 Kbytes of internal RAM occupying the first 4 K, with the on-board static RAM residing in the next 32 KBytes of address space. The B401 is a size 1 TRAM consuming 1.5 W.

The limited memory available on this TRAM makes it most suitable for signal processing, feature extraction and systolic processing. This TRAM offers a low cost entry to 32-bit parallel computing, with the option of employing on-chip floating point calculations by choosing this TRAM with the T800 installed.

Ordering details

Company	*Order No.*	*Description*
Inmos	IMS B401-5	32 KBytes SRAM – IMS T800-25
Inmos	IMS B401-8	32 KBytes SRAM – IMS T425-25
Transtech	TTM2-8	128 KBytes SRAM – IMS T800-20
Transtech	TTM2-42	128 KBytes SRAM – IMS T425-20

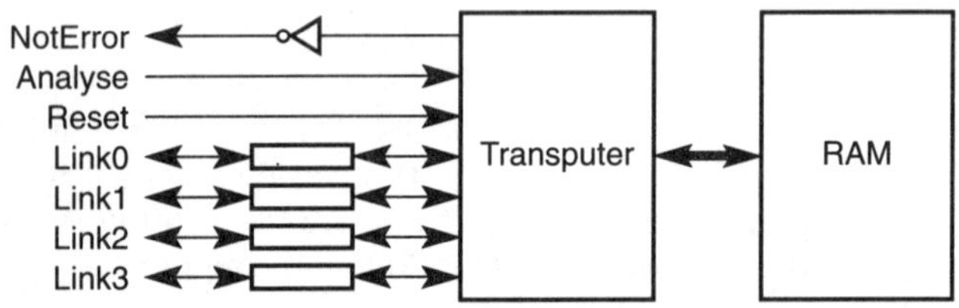

Figure 5.1 General TRAM with multiplexed access to RAM.

5.3.4 IMS B410 TRAM

The B410 TRAM utilizes the IMS T801 transputer which, with demultiplexed address and data buses, allows rapid execution to take place when employed with 100 ns static RAM (Figure 5.2). The T801-20 executes with a cycle time of 50 ns, achieving a maximum throughput of 20 MIPS peak and 10 MIPS sustained. The address space of the B410 is 164 KBytes in total, which is comprised of 4 Kbytes of internal RAM and 160 KBytes of external RAM. The pin designations for the B410 TRAM is the standard for a size 2 TRAM; this expresses the active pins on the 16 pins which are indicated by the index triangle. The remaining set of pins can be used to stack another TRAM on top. The power consumption of this TRAM is 3 W.

This TRAM enables all of the usual applications, requiring high speed and floating point calculations, to be hosted. However, the limited memory available on this TRAM does place restrictions on its uses.

Ordering details

Company	*Order No.*	*Description*
Inmos	IMS B410-5	160 KBytes SRAM – IMS T801-20
Transtech	TTM30	256 KBytes SRAM – IMS T801-25

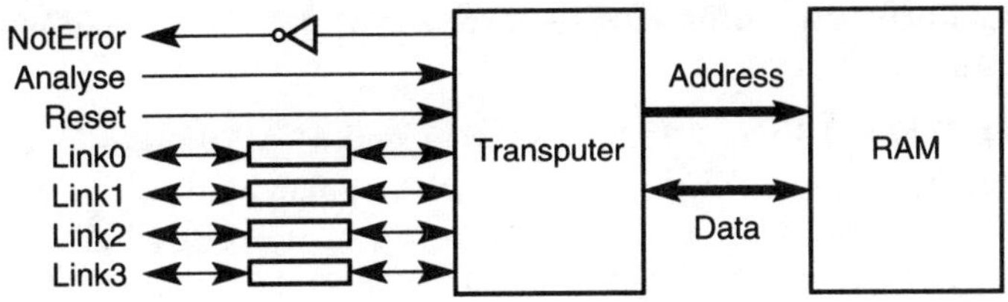

Figure 5.2 General TRAM with demultiplexed access to RAM.

5.3.5 DRAM-based multiplexed address and data bus TRAMs

These TRAMs, of which the B411, B426, TTM3 and TTM16 are examples, are general purpose TRAMs with multiplexed data and address buses. They are based upon the T800 which can be obtained at a wide range of speeds from 17.5 MHz to 30 MHz, achieving up to 30 MIPS peak, and can support quantities of memory from 1 Mbyte to 4 MByte of RAM along with the on-chip 4 KByte internal RAM. These TRAMs are of the type indicated in Figure 5.1, and are presented as a size 1 TRAM. The power consumption can be up to 4 W, depending upon the size of the memory employed.

These TRAMs are suitable for a wide range of applications requiring speed and large quantities of RAM. They can also exploit the compatibility between the pin-outs of the T425 and T800, and as such can have either processor installed. However, for floating point calculations the T800 is necessary.

Ordering details

Company	*Order No.*	*Description*
Inmos	IMS B411-5	1 MByte DRAM – IMS T800-25
Inmos	IMS B411-8	1 MByte DRAM – IMS T425-25
Inmos	IMS B426-5	4 MBytes DRAM – IMS T800-25
Transtech	TTM3	1 MByte DRAM – IMS T800-25
Transtech	TTM16	4 MBytes DRAM – IMS T800-25

5.3.6 DRAM-based multiplexed address and data bus TRAMs with subsystem control

The B404, B427, TTM6, TTM17, TTM18 and TTM19 are all examples of these general purpose TRAMs with multiplexed data and address buses. They are based around the T800 or T425, which can be obtained at a wide range of speeds from 17.5 MHz to 30 MHz and can support quantities of memory from 1 Mbyte to 16 MByte of RAM, along with the on-chip 4 KByte of internal RAM. These are of the type indicated in Figure 5.3. They have the ability to implement full subsystem control to enable the TRAM to reset or analyse a subsystem or other TRAMs. These TRAMs range in dimension from size 1 to size 3, and power consumption can be up to 6 W.

These TRAMs are suitable for a wide range of applications requiring speed and large quantities of RAM. They can also exploit the compatibility between the pin-outs of the T425 and the T800, and as such can have either processor installed. However, for floating point calculations the T800 is necessary. The largest 16 MByte would be necessary in circumstances where it was impracticable to split the memory across multiple processor elements.

Ordering details

Company	*Order No.*	*Description*
Inmos	IMS B404-5	2 MBytes DRAM – IMS T800-25
Inmos	IMS B404-8	2 MBytes DRAM – IMS T425-25
Inmos	IMS B427-5	8 MBytes DRAM – IMS T800-25
Transtech	TTM6	2 MBytes DRAM – IMS T800-25
Transtech	TTM17	4 MBytes DRAM – IMS T800-25
Transtech	TTM18	8 MBytes DRAM – IMS T800-25
Transtech	TTM19	16 MBytes DRAM – IMS T800-25
Mark Ware Associates	TM8022S-4	8 MBytes DRAM – IMS T800-25

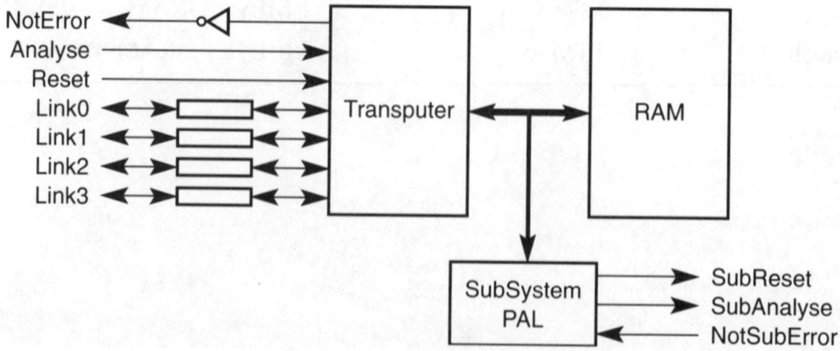

Figure 5.3 General TRAM subsystem control multiplexed access to RAM.

5.3.7 DRAM-based demultiplexed address and data bus TRAMs with subsystem control

The T801 transputer is the processing element in these TRAMS. Its demultiplexed address and data buses enable fast, 2-cycle accesses to external memory, as displayed in Figure 5.4. Examples of these TRAMs are noted below, and are produced with varying DRAM configurations. The Transtech devices have additionally 32 KBytes of 2-cycle SRAM and page control, providing a speed increase. These TRAMs have the ability to implement full subsystem control to enable the TRAM to reset or analyse a subsystem or other TRAMs. These are of size 2 TRAM, and exhibit a power consumption up to 6.5 W.

These TRAMs can be used to provide extremely fast computer systems requiring large quantities of memory. They can be used as a master controller by using the subsystem control features of the TRAM.

Ordering details

Company	*Order No.*	*Description*
Inmos	IMS B428-12	2 MBytes DRAM – IMS T801-25
Transtech	TTM32	2 MBytes DRAM – IMS T801-25
Transtech	TTM34	4 MBytes DRAM – IMS T801-25
Transtech	TTM38	8 MBytes DRAM – IMS T801-25

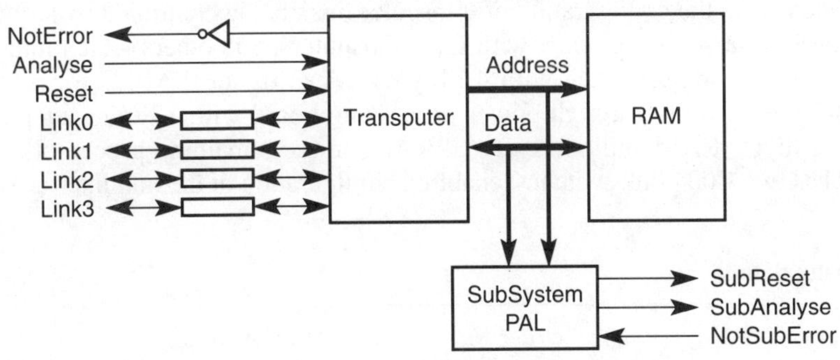

Figure 5.4 General TRAM subsystem control demultiplexed access to RAM.

5.3.8 Multiple transputer boards

TRAMs

There are some applications where it necessary to have multiple high power processing elements with minimum memory, for example image processing and digital signal processing. These can be achieved by employing single TRAM multiple transputer arrays, for example the TM4x8002-2 which integrates four T800s onto a size 2 TRAM. The array is organized as a square, with each T800 having one link connected to each of its two nearest neighbours. The remaining two links of each T800 are brought to the TRAM pins. They are organized as a group for reset and analyse, whilst the four noterrors are ORed together before being expressed at the TRAM pin.

Ordering details

Company	*Order No.*	*Description*
Mark Ware Associates	TM4x8002	4 × T800-25

Non TRAMs

Certain manufacturers build multiple transputers on boards to be built into their own systems. These tend to be at the large system end of the market, and so would not be installed by a system builder. However, there are some companies who offer multiple transputer boards to be integrated into smaller host systems. Microway market the Quadputer, featuring four T800-25 with up to 4 Mbytes of RAM per transputer. This is an AT form factor board utilizing the Inmos standard B004 interface. The same company's Linkputer allows eight Quadputers to be connected, with the configuration of transputer links being controlled by software. Sension have a similar product with four transputers, and others which integrate up to nine transputers, each with 4 Mbytes RAM, on an IBM AT form factor board. They also offer a high power processing board with 17 transputers, one with 4 Mbytes RAM and non-external RAM on the remaining processors. This board has two C004 link switches, enabling configuration of the link interfaces.

Ordering details

Company	*Order No.*	*Description*
Microway	Quadputer	4 × T800-25 4 × 4 Mbytes RAM
Sension	Fast4-T8	4 × T800-20 4 × 1 Mbyte RAM
Sension	Fast9/36M	9 × T800-20 9 × 4 Mbytes RAM
Sension	Fast17-T8	1 × T800-20 4 Mbytes RAM with 16 × T800-20 no RAM

5.4 Standard interface TRAMs

Interface standards

These have arisen with the developing requirement for communications between computers, instruments and mass storage devices. This is accomplished by interconnecting these devices to a common bus. This means each device must be interfaced so that its logic convention, signal levels, timing protocols, etc. are compatible with those of the other devices sharing the bus.

Traditionally, the approach to interface design has been to provide each device with a set of specialized control, data and status signals. A design readily provided direct control of a device's functionality, but was too dependent on the decisions made by the interface engineer, leading to there being as many interfacing techniques as there were design engineers. Thus not only were interfaces from different manufacturers incompatible, but even devices from the same manufacturer were incompatible. To accommodate this variety (logic levels, timing, etc.), many different adaptors had to be built, creating a mass of interconnections. As the complexity of devices increased, so did the complexity of the adaptors. The resulting interfacing costs became unmanageable and software costs rose dramatically. The day of the interface standard had surely dawned. The first widely accepted standard was that designed by Hewlett Packard called the HPIB (Hewlett Packard Interface Bus), which became standardized into the GPIB or IEEE-488. This was followed by the SCSI (Small Computer Systems Interface), having its roots in Shugart Associates, a hard disk manufacturer. These bus-based interfaces have gained wide acceptance, and devices that claim to be compatible usually behave properly. With the advent of Local Area Networks (LANs), much activity has proceeded in the direction of interconnecting computer systems. This is commonly realized by Ethernet.

Need for interface-specific TRAMs

The usual interface supported by transputers is via their four links. However, these have the limitation that they will only connect to other devices with links. There are interface standards which have developed, allowing diverse equipment to intercommunicate. These generally were originated by computer or peripheral builders, and have become industry standards. There is a significant advantage for a designer of new equipment to adopt these standards, as they readily enable interfacing to a wide variety of peripherals.

Generalities about interface-specific TRAMs

These TRAMs can be considered to comprise two sections. The first is a typical transputer-based module which communicates with other TRAMs via links in the usual way. The second section is interface-specific, and is interfaced to the buses of the transputer. This enables a TRAM of this nature to control the interface in a standard manner, and also to present a familiar interface to other

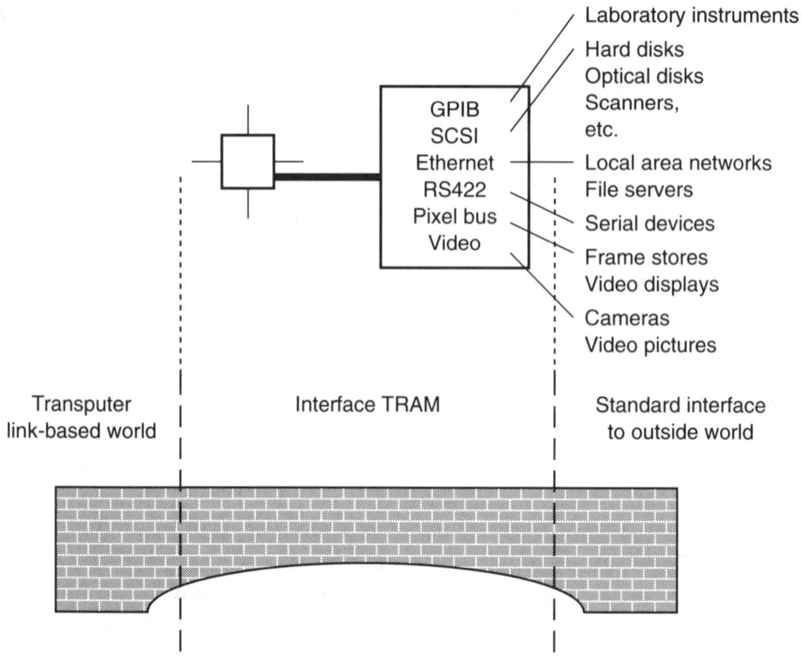

Figure 5.5 The interface bridge.

TRAMs. This can be visualized as the interface bridge presented in Figure 5.5. This approach has been necessary to allow the closed inter-transputer world to make use of equipment which has been developed for other computer systems, giving rise to extensive savings in hardware development times and, ultimately, costs. Using this approach, transputer-based systems can readily make use of a wide range of devices, from video or hard disks to scanners.

These devices have been developed utilizing existing standards and support communications, of various bandwidths, with a wide range of media. To interface to other TRAMs the user plugs in the TRAM in the usual way and establishes the appropriate links. The TRAM presents the standard interface on familiar sockets, and the bus is connected to make the electrical connection. The interface software executes on the transputer and provides applications with the ability to communicate with the external device.

5.4.1 IMS B421 IEEE-488 GPIB TRAM

Introduction to the general purpose interface bus IEEE-488

This standard, originated by Hewlett Packard, was eventually standardized in the documents IEEE-488.1 1987 and IEEE-488.2 1987. The purpose of the standard is to allow the interconnection of programmable instruments with minimal engineering effort. Configurations using various instruments produced by different manufacturers may be established. The standard provides a set of rules providing a communications link between independently designed products. The standard defines the electrical, mechanical and functional characteristics of the interface. The data exchange between the interconnected equipment is digital, with a maximum rate of 1 Mbyte/second, the maximum number of devices is 15, and the total transmission path should not exceed 20 metres.

Overview of IEEE-488 standard

The GPIB, shown in Figure 5.6, is a parallel bus which interconnects two or more instruments. It is presented on an Amphenol 24-way connector using shielded cable. An instrument exhibits two types of functionality: interface functionality determining the passing of messages to and from the instrument; and device functionality determining the specific instrument control.

The GPIB is made up of 16 transmission lines which have the following attributes:

- eight data lines;
- three handshake lines for control of data transfer;
- five general interface management lines.

The data is transferred from the talker to the listener, and commands passed from the controller to devices. Transmission is asynchronous, and is normally controlled by a 3-wire handshake that enables the talker/controller to synchronize its readiness to transmit to the listener's readiness to receive data.

A controller directs the flow of data by assigning devices to being listeners or talkers. Each device is assigned a unique address, via dip switches, which is used by the controller for configuration purposes.

Interface specification

Interconnected devices	Up to 15 maximum on one bus
Signal lines	Star or linear bus network 20 metres max. transmission path length
Connectors	24-way Amphenol/Cinch S7 or CHAMP
Message transfer scheme	Byte-serial bit-parallel asynchronous transfer using interlocked three wire handshake
Data rate	1 MByte/s maximum over limited distances Typical 500 Kbytes over max. path length
Address capability	Primary addresses, 31 talk, 31 listen Secondary addresses, 961 talk, 961 listen
Control shift	Can be delegated, never assumed Maximum of one talker and 14 listeners.

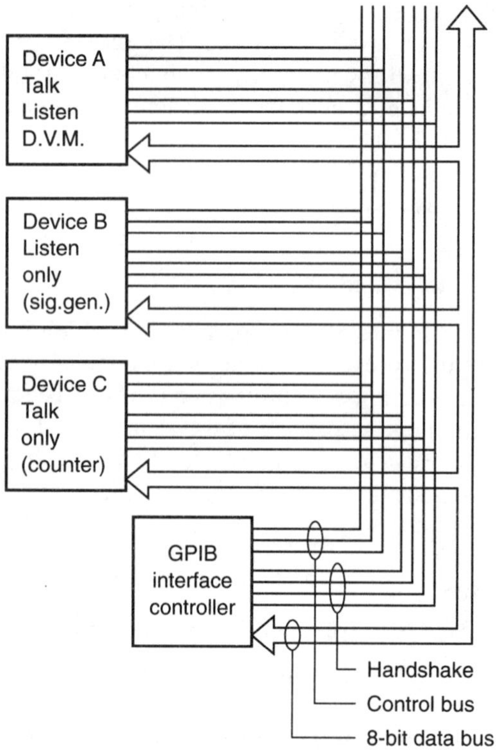

Figure 5.6 The GPIB interface bus structure.

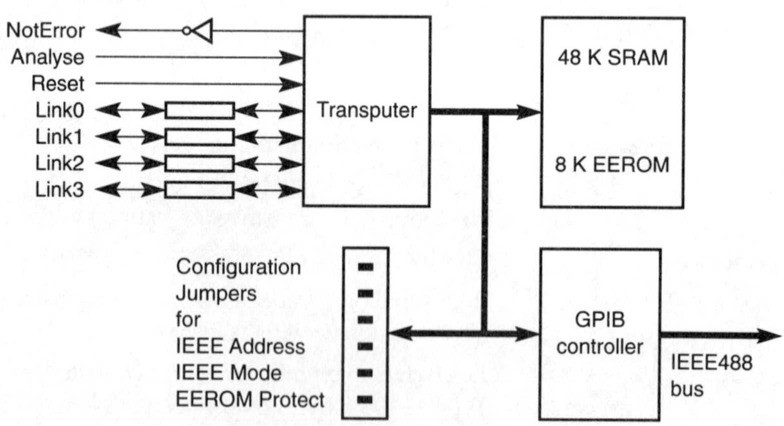

Figure 5.7 IEEE-488 GPIB TRAM.

IMS B421

The IMS B421 provides connection to the IEEE-488 bus via the standard 24-way connector. The supplied software IMS F001 provides access to the facilities of the TRAM via library calls, freeing the user from concerns about low level hardware details and minimizing the development of software by the programmer. The software communicates using an input channel and an output channel, which can be mapped onto the links of the TRAM, or alternatively implemented as soft links to a user program executing concurrently on the local transputer (Figure 5.7). It is necessary for the successful implementation of IEEE-488.2 that a device accepts, retains and returns various items of identifying information upon demand. These are stored in an EEROM by IMS F001 commands (incidentally, at 8 K the EEROM has much more capacity than required for this purpose, so the remainder can be used for other purposes as determined by the user). However, this use should be treated with some care, as the EEROM can only accept a maximum of 10,000 writes per cell. There is a physical jumper which enables the user to protect against writing into the EEROM.

Application note This device could be used as part of a larger computer system which will analyse data received across an IEEE-488 bus that is connected to a series of instruments. Alternatively, this TRAM could be built into a new instrument providing the IEEE-488 interface to an existing network and simultaneously offering extensive control of the instrument and powerful processing of the locally collected data.

Pin-out The pin designations for the IMS B421 TRAM is the standard for a size 4 TRAM supporting the usual four links.

External connections The TRAM has nine jumpers which enable the user to set the IEEE-488 device number of the TRAM, and to determine its mode of operation, that is: talk, talk and listen, listen or controller.

The IEEE-488 is presented on a 26-way header, two rows of 13 pins on 0.1" pitch. A ribbon cable is required which will connect to the header and offer the standard IEEE-488 connection to the other devices. It is best that this cable be kept as short as possible, and the IEEE-488 connector be mounted on a panel of the computer enclosure.

Processor This TRAM is based on the T222 transputer in conjunction with a TMS9914A GPIB controller chip.

Memory The TRAM is supplied with 48 Kbytes of external RAM, 4 Kbytes internal RAM and 8 Kbytes of EEROM; the nine jumpers and the GPIB control registers are memory mapped.

Power requirements

Supply voltage:	4.75–5.25 V.
Power consumption:	7 W.

Physical dimensions

Industry standard Size 4 TRAM.

Component height above pcb: 9.2 mm.

Component height below pcb: 3.7 mm.

Ordering details

Company	*Order No.*	*Description*
Inmos	IMS B421-10	GPIB TRAM–IMS T222-20
Inmos	IMS F001B	GPIB software libraries

5.4.2 IMS B422 SCSI TRAM

Introduction to the SCSI

The SCSI (Small Computer Systems Interface) standard has been developed from the SASI (Shugart Associates Systems Interface), and is generally referred to as the 'scuzzy' bus by the computer community. This is a systems-level interface which enables up to eight controllers to share a SCSI bus, with one of these controllers being the host adaptor. Many peripherals are now being made with SCSI interfaces which enable hardware manufacturers to offer their wares to computer owners using IBM PC, Apple Macintosh and larger systems. Some difficulties have arisen in the past as different manufacturers have implemented slightly different 'flavours' of the standard. However, no difficulties should arise if peripherals conform to ANSI X3,131-1986 specifications. This standard defines the electrical, mechanical and functional characteristics of the interface. The data exchange between the interconnected devices is digital, with a maximum data rate of 4 Mbytes/second, the maximum number of device controllers is eight, and the total transmission path length should not exceed six metres.

Overview of SCSI standard

The SCSI is a parallel bus which interconnects two or more SCSI devices. It is presented on a 50-way connector, two rows of 25 pins on a 0.1" pitch. This interface lends itself to connecting mass-storage devices to computer systems. However, many other peripherals are produced which make use of this standard. A complete list of peripherals is not realistic, but would include the following: hard disks, optical disks, tape streamers, CD ROMs, printers and plotters. This interface had been adopted by Apple for the original Macintosh to remove the difficulty of this computer being a closed system with no expansion slots. Other companies – Future Domain, Western Digital and Adaptec – have developed SCSI host adaptors for the IBM AT/PS2 range of computers. This activity, along with that occurring with larger computer manufacturers, has led to the rapid development of equipment with SCSI interfaces. The size of this market has driven the price of peripherals down, and further increased the popularity of this interface.

The SCSI bus is implemented on a 50-way ribbon cable with an IDC connector, and comprises 18 transmission lines which have the following attributes:

- eight data lines and one parity line;
- eight bus management lines.

Commands are issued by the host adaptor and executed by the target. The target controls the transfers on the bus to the destination. Each SCSI device has a site for termination resistors and has jumpers or dip switches to select its address.

Interface specification

Interconnected devices	Up to eight maximum on one bus
Signal lines	Linear bus network Six metres max. transmission path length Transmission line terminated at both ends (Figure 5.8).
Connectors	50-way IDC/ribbon cable or 25-way 'D'/shielded round cable
Message transfer scheme	Byte-serial bit-parallel synchronous transfer
Data rate	4 MBytes/second maximum at limited distances 1.5 Mbytes/second sustained
Address capability	Targets 0–7
Control	Maximum of 1 initiator and 1 target at a time

IMS B422

The IMS B422 enables interfacing to the SCSI bus via the standard 50-way connector (Figure 5.9). The IMS F002B SCSI software library supplied with the TRAM provides a low level interface between an application and peripherals. This presents the user with a consistent view of the hardware even when additions or replacements are necessary. This TRAM enables a system to be readily interfaced to a vast range of different mass-storage and other peripherals.

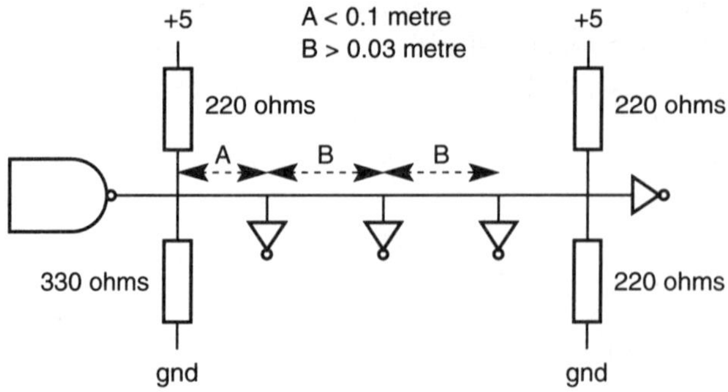

Figure 5.8 SCSI termination and connection.

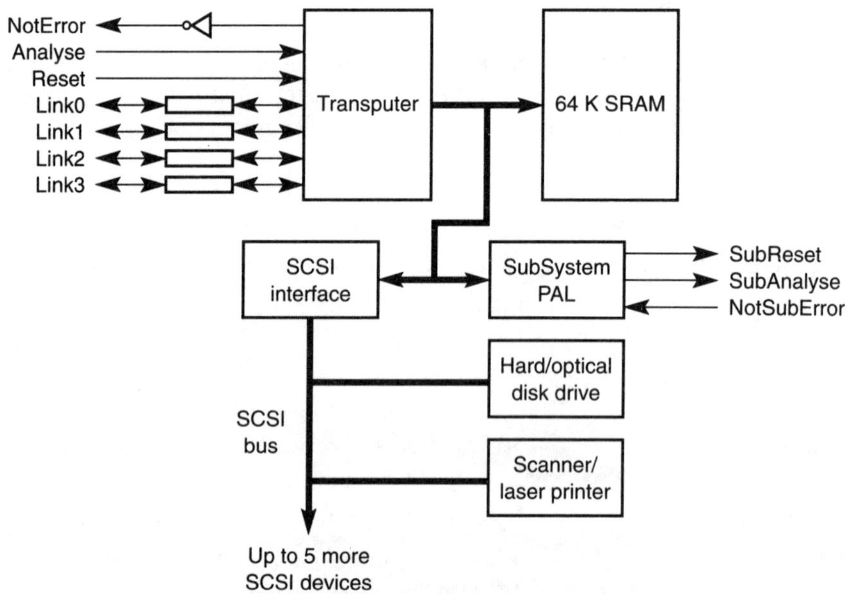

Figure 5.9 SCSI TRAM with devices.

Application note This device, in initiator mode, is extremely useful when wishing to make connection from a transputer system to large disk drives or other mass-storage media. It could also, in target mode, be useful as part of an embedded controller in a laser printer where large quantities of data must be transferred rapidly. The software supplied enables both of the modes to be determined by the system builder. Another attraction of SCSI is that the cable connections are readily made.

Pin-out The pin designations for the IMS B422 TRAM is the standard for a size 2 TRAM supporting the usual four links.

External connections The SCSI is presented in the standard manner, on a 50-way header, two rows of 25 pins on 0.1" pitch on a 50-way 0.05" pitch ribbon cable. If the SCSI bus exits from the enclosure, a short connecting cable should be made with a panel mounting socket fixed to the outside of the system. Each end of the SCSI bus should be terminated. It must be borne in mind that this bus is treated as a group of terminated, lossless transmission lines of characteristic impedance of 100 ohms. The bus will exhibit losses and impedance mismatches due to multiple taps to the transmission line. Usually, the TRAM with its termination is at one end with a terminated peripheral at the remote end. Overall, the bus should be kept as short as feasible and the loads evenly separated while not being closer than 0.03 metres and stub length no greater than 0.1 metre. If required, the B422 can supply power for termination purposes via one of the conductors and the on-board termination SIL resistor packs can be removed.

Processor The B422 is based on the T222 transputer interfaced to a SCSI interface device.

Memory The B422 is supplied with 64 Kbytes of external SRAM, 4 Kbytes internal RAM with SCSI registers being memory mapped.

Power requirements

Supply voltage: 4.75–5.25 V.
Power consumption: 5 W.

Physical dimensions

Industry standard size 2 TRAM
Component height above pcb: 13.9 mm.
Component height below pcb: 4.1 mm.

Ordering details

Company	*Order No.*	*Description*
Inmos	IMS B422-10	SCSI TRAM – IMS T222-20
Inmos	IMS F002B-1	SCSI software libraries

5.4.3 IMS B300 Ethernet connection

Introduction to Ethernet IEEE 802.3

The Ethernet LAN was developed by Xerox (Ethernet, 1976), DEC and Intel during the 1970s, and was specified in 1980 (The Ethernet, 1980) as baseband utilizing CSMA/CD. This was then adopted as the IEEE 802.3 standard. A LAN of this nature enables the local interconnection of computers and peripherals typically within a single building, enabling users to share the resources while minimizing the quantity of cabling required. The standard defines the electrical, mechanical and access technique of the network. The data exchange across the network proceeds at 10 Mbits/second, the maximum number of stations per segment is 100, and the maximum transmission path length of a segment is 500 metres (Figure 5.10).

Overview of IEEE 802.3

The Ethernet medium is a passive coaxial cable onto which the signal is directly modulated. Stations are connected to the medium via an interface cable which taps into the coaxial cable. The transceiver must be adjacent to the cable tap, but the interface cable may be up to 50 metres in length. Each segment can be up to 500 metres long; the cable is marked at 2.5 metre intervals so that taps can be made avoiding unwanted reflections. Segments can be linked together using repeaters attached at any transceiver point. A point-to-point link of up to 1000 metres may also be included in the configuration, enabling the linking of segments in different buildings. Each station must have a path to every other station which does not contain more than two repeaters, thus limiting the overall distance between the most remote stations to 2.5 km. This maximum distance limits the round-trip propagation delay time of signals across the network.

The transfer of data across the network proceeds using the CSMA/CD access method. All stations are peers, and can attempt to transmit data at will (Multiple Access – MA). When a station wishes to transmit to a station it first acquires the network, ensuring that no other station is transmitting data by listening to the medium (Carrier Sense – CS). When the station perceives that the medium is free it starts to transmit, and the signal propagates throughout the medium; however, there is the possibility that another station has simultaneously reached the same state and that a collision will occur. A collision is detected by the transmitting station comparing the transmitted signal with the received signal and flagging a collision should a discrepancy arise (Collision Detect – CD). The station will then send a jamming signal to ensure other stations are aware of the collision. The station will attempt to transmit again at a later time, determined by a binary exponential back-off algorithm. The probability of collision is minimized by the limitation of the physical extent of the medium, thereby reducing the time available for a collision to occur. Each station may make a total of 16 failed attempts before the transmission is abandoned. Heavy traffic on the network will give rise to an increasing number of collisions and a reduction in overall network data transfer rate.

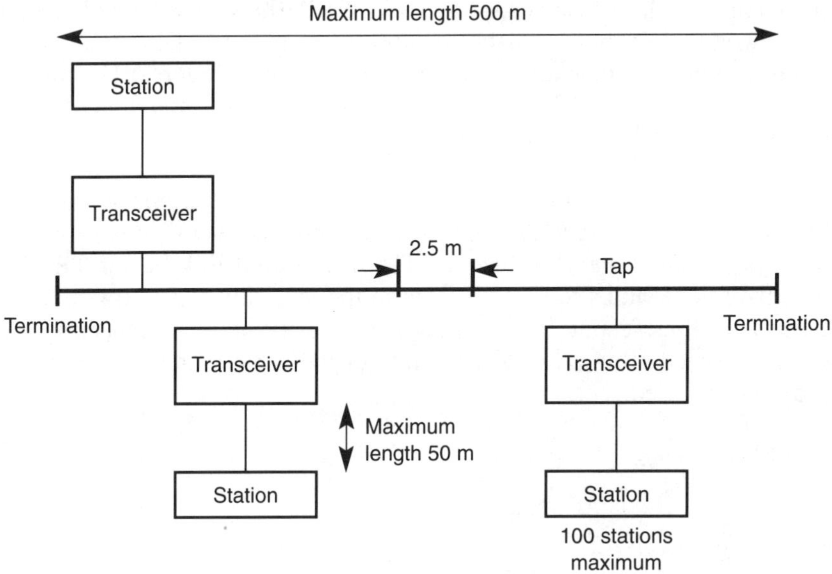

Figure 5.10 Single Ethernet segment.

Data is transferred in packets each containing destination and source addresses (12 octets), data (46–1500 octets) and error checking (four octets). The data can be transmitted to a single station or groups of stations as determined by the destination address. There is no restriction on the sequence of octet values appearing in the data field.

Interface specification

Interconnected devices	Up to 100 maximum on one segment
Connectors	IEEE 802.3 AUI
Signal lines	Non-rooted tree network 2.5 km max. transmission path length
Message transfer scheme	Bit-parallel synchronous with CSMA/CD
Data rate	10 MBits/second maximum, load dependent
Address capability	47-bit, all Ethernet addresses are unique
Control shift	Maximum of one talker at a time

IMS B300 The IMS B300 is a self-contained and cased unit with four transputer links providing connection to transputer-based networks across an Ethernet network. Host computers executing suitable software with TCP/IP can access the links over the network. This unit enables the design of a physically distributed network of transputer systems by providing transparent access to links within each system. Transputers can be connected to networks either by installing a motherboard in a host machine, or by using the B300 with a stand-alone transputer network. Software is available to allow B300, IBM PC, DEC VAX and Sun systems to support remote access to transputer networks (Figure 5.11). Since the B300 provides IEEE 802.3 AUI connection, various IEEE 802.3 media can be coupled into using the appropriate taps.

Application note The links as presented by the B300 are buffered to EIA RS422 levels, differentially driven to reduce noise and maintain isolation. The target system must therefore use a B415 differential link buffer TRAM to enable the remote links to be connected to the local links. The B415 is capable of 20 Mbits/second link operation so long as the interconnecting cable does not execeed 10 metres; as the length of the cable increases, slower link speeds must be used.

Power requirements

Supply: AC mains.
Power consumption: 25 W.

Ordering details

Company	*Order No.*	*Description*
Inmos	IMS B300-1	B300 network connection device
Inmos	IMS S707	Network support software IBM PC
Inmos	IMS S607	Network support software VAX VMS
Inmos	IMS S507	Network support software Sun3 and Sun4

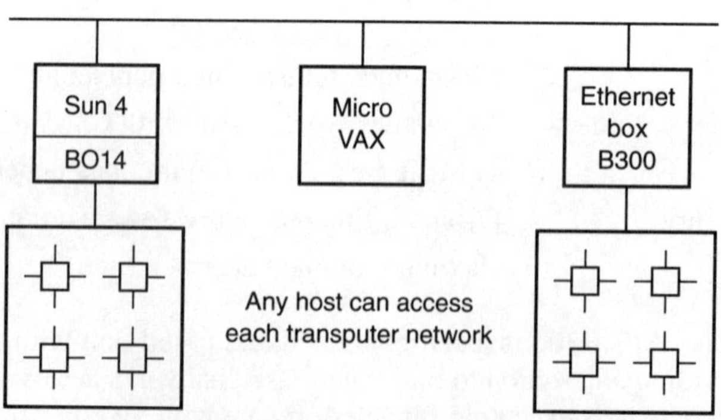

Figure 5.11 Mixed network over Ethernet.

5.5 Image processing systems

5.5.1 Introduction

Image processing applications tend to give rise to large quantities of data, and many operations to be executed at high speed. The transputer, with its processing power and ability to exploit parallelism, has been applied to this field with some vigour by a range of companies. There is a rapid rate of development in this area, and latest product information should be sought before a purchase is made. Image processing systems, depending upon the application, can be thought of as comprising some or all of the following components: illumination, video camera, frame grabber, real-time image processing, frame storage, frame processing, display driver and display (Figure 5.12). The devices at each end of the list are not specific to transputer systems, and despite seeming commonplace are anything but simple.

Illumination of objects that are to be imaged can present difficulties, most particularly that of achieving even illumination of metallic or highly reflecting objects. Short pulses of light, or electronic optical shutters, can be used to 'freeze' objects in motion so that the limitation of 1/25 second exposure is overcome. Video camera technology has benefited from semiconductor advances and the consumer boom in portable low power domestic video cameras. Charge Coupled Devices (CCD) cameras are now available at a low price, and have the added advantage that they integrate the accumulated charge, allowing for very short to very long exposures. The display for output purposes can be high resolution monitors or, in some cases, the video display of the host computer.

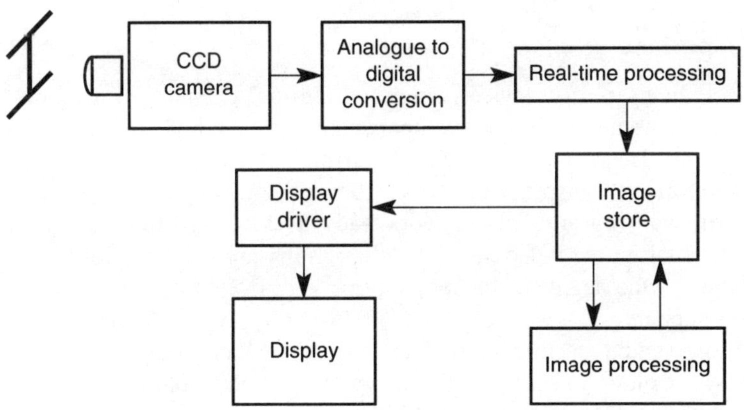

Figure 5.12 Components of an IP system.

5.5.2 Transputer-based frame store and display driver

Inmos produce these components as separate TRAMs, B408 and B409. The frame store TRAM B408 consists of a T800 and 1 Mbyte workspace. Additionally, there is a 1.25 Mbyte dual port RAM which can be accessed by both the T800 and an address generator, with the output data being made available to a pixel bus. The T800 is available for the processing of images in its workspace and transferring the image to be displayed to the dual ported RAM. The address generator runs autonomously of the processor, and can generate pixel data transmission up to 100 Mbytes/second. This is capable of supporting displays up to 1024 × 768 pixels. The pixel bus is presented on a 60-way ribbon cable with IDC connectors. It has a 32-bit data path and control signals to synchronize the transfer of pixel data. The bus is designed so that multiple B408s can transmit data in a synchronized manner. The display TRAM B409 receives the pixel data from the pixel bus and routes it to the CLUTs (Colour Look Up Tables). The pixel data is used as an address by the CLUT whose output is applied to three DACs (Digital to Analogue Converters), one for each of red, green and blue. The CLUT has 256 memory locations, each of 18 bits, therefore 8 bits of pixel data will produce 3 × 6 bits of DAC data, therefore enabling the display to produce 256 colours selectable from a palette of 262,144 (2^{18}). The B409 has three channels and three CLUTs, and sufficient flexibility to support either one or multiple displays. The contents of the CLUTs can be read or modified by the on-board transputer T222. Each of the pixel channels has RGB outputs presented on SMB connectors, and are capable of driving standard 1 V peak to peak into a 75 ohm load. Both the B408 and B409 are size 8 TRAMs.

5.5.3 Integrated graphics display devices

There is a wide series of devices which fall into this category, which consists of transputers combined with other components to provide high speed, high resolution graphics. These boards all use T800 transputer links as their connections to the supporting transputer network. They generally consist of a T800 processor with varying amounts of workspace RAM (2–8 Mbytes) and up to 2 Mbytes of dual ported video RAM. All have software available for them which will enable a wide degree of functionality ranging from graphics functions to X-Windows support.

Inmos and Transtech both offer a TRAM which integrates a T800 with G300 CVC (Colour Video Controller) to provide high resolution graphics. Each has 2 Mbytes of dual ported video RAM, allowing for resolutions up to 1280 × 1024 pixels, 8-bits/pixel. The B419 from Inmos, a size 6 TRAM, has 2 Mbytes of workspace RAM, while the TTG3 from Transtech, a size 4 TRAM, can have up to 8 Mbytes of workspace RAM. A similar product from AIKI, the G8, a size 4 TRAM, is capable of stereo graphics using an appropriate display. Sension have a product, GPM96, which employs the Motorola 96002. This processor supports the high throughput required of high-speed graphics systems, and controls all of the timing and display via software. The GPM96 has 4 Mbytes of RAM which is private to the T800, with a further 4 Mbytes which is either shared or private to the 96002. Two 1 Mbyte frame stores are provided for the 96002, either of which can be displayed by software. These four devices produce RGB video signals which can drive standard 75 ohm cables directly. In cases where the synchronizing video signals are required, the B419 and GPM96 also produce separate video sync signals.

5.5.4 Transputer-based frame grabbers

Frame grabbers are devices which enable a video signal, typically from a video camera, to be digitized and stored pending further manipulation. Some of those presented here actually do more than just grab a video frame, but also enable front-end processing and the display of captured images. Transtech supply the TTG-F which has a T800, 4 Mbytes of workspace RAM and 1 Mbyte of video RAM, enabling frames of 1024 × 1024 pixels, 8-bits/pixel. This TRAM can also be used to display the live digitized video signal. The Harlequin, marketed by Sension, captures video signals at a resolution of 512 × 512 pixels, 8-bits/ pixel. This is an IBM AT board and has two display buffers, either of which can be displayed during frame acquisition. The board is T800 based, has 1 Mbyte workspace RAM, 0.5 Mbyte of video RAM, and presents the transputer links. The B429 video image processing TRAM enables front-end image processing of video data as it is being captured. This TRAM, with a T800, 2 × A110, 1 Mbyte of workspace RAM and 3 × 0.5 Mbyte frame buffers, can acquire images of 512 × 512 pixels, 8-bits/pixel. The A110s, image and signal processing subsystems, can perform convolution filtering of the captured data. Sension also offer the IP1, which is a frame grabber combined with a 2D Convolver PDSP16488. This also can acquire video signals from four sources and display real-time captured data. Another IBM AT board utilizing the T800 is the Microeye TM/TC from Digithurst. This device can grab full-colour frames into a picture store and display frames held in a separate frame store.

These devices capture frames of video data, PAL or RGB if appropriate, usually generated by television cameras. The on-board T800 can be employed in feature extraction, internal routing of data and the provision of links to further transputers. It should be noted that the transfer of video frames, up to 4 Mbits of data/frame, makes severe demands upon links if the entire frame has to be distributed to a series of transputers. For real-time processing of images, workers at Liverpool University (Professor J. Lucas *et al.*, Department of Electrical and Electronic Engineering) have developed a high-bandwidth dedicated image communication bus freeing the transputers from image transfer, thereby circumventing the bottleneck imposed by data transfer over links.

5.6 Motherboards and interfaces

5.6.1 Motherboards and standard interface boards

The concept of a motherboard is used to exploit the developments made by other computer manufacturers. This enables a user to employ commonly available computer systems to front-end a transputer implementation. Advantages accrue from this approach, as it does away with the restriction that the transputer application must be developed on a wholly dedicated transputer system. The motherboard can electrically interface at the bus level of the host computer. The intercommunication between the host computer and the transputer is either at link level or by shared memory. If greater processing power is required, a motherboard, once fully populated and installed in a host computer, can give access to further transputer networks via the links. Motherboards exist for many common computer systems, allowing a low cost entry into parallel processing. This has contributed to the uptake of transputers in academic and development environments.

The organization of motherboards can vary considerably, some having a fixed pipeline of links, using two links from each transputer with the remainder connected to a crossbar switch or presented on an external connector. Others have all their links connected to a crossbar switch, enabling software configuration of the network. Further configuration is available on boards where the uncommitted links are brought to the edge of the board, enabling hardware connections to be made.

Archipel, Inmos, Meiko, Parsytec and Transtech all provide motherboards for some or all of the following host computers, although each offering is not identical: IBM PC and compatibles; IBM PS/2 range; NEC 9800 series PC; VME slave Sun3/4 systems; Sbus for Sun SPARCstation; Apollo; and Apple Mac II. Other companies offering motherboards for computer systems are: Paratech Solutions (TMac: Apple Macintosh via SCSI port), San Diego Levco (Translink: Apple Mac II via Nubus slots), Perihelion (ATW: Atari Mega ST) and Linkspeed Technology (BLINK: BBC 1 MHz Expansion Bus).

5.6.2 Interfaces

Many of the companies mentioned in this text can supply interfaces, including interfaces for analogue to digital conversion, digital to analogue conversion and optical fibre connections. It is even possible to acquire a silicon hard disk, that is, a transputer-controlled memory array which can be used instead of a hard disk.

It is possible to make your own by using the IMS B430 prototyping TRAM which has an area of the board onto which you build your circuit. Flight Electronics also produce a prototyping board which has a breadboard area available for non-solder connections.

5.7 Large transputer configurations

Applications that are numerically intensive (CFD, molecular modelling, etc.) require scalable computer solutions demanding the use of considerable quantities of transputers or other high performance processing elements. A number of companies offer solutions to this problem, Meiko, Inmos, Archipel, Sension, Parsys and Parsytec included. Systems of 64 and more transputers can be readily obtained which can provide computing power of supercomputer performance. Careful inspection of the interconnection capability of transputer links should be made to ensure that the system can be configured to suit the application. A detailed description of this range of systems is beyond the scope of this chapter. All of the companies mentioned below can offer an extensive range of storage and I/O elements which can be used to provide a wide degree of functionality. These include elements for frame grabbing, display, octal serial ports and hard disk storage via SCSI interface.

5.7.1 Meiko Computing Surface

An example of a powerful parallel environment is the Computing Surface from Meiko. This system is comprised of a series of computing elements connected by transputer links to the CSN (Computing Surface Network). The processing elements can be either transputers or other high performance processors using transputers for communications access to the CSN. The number of transputers required depends upon the processor element's requirements, thus the i860 element will use two T800s as NIPs (Network Interface Processor) while the SPARC element requires four T800s as NIPs. The Computing Surface has an additional interface to all elements, called the **supervisor bus**; this provides for the reporting of hardware and software failure, and the electronic configuration of the elements. This enables the module host to configure the topology of the network of elements. The Computing Surface can be divided into a number of virtual machines, allowing users to have protected access to the resources allocated to them. This implements a multi-user service without the drawback of task swapping on individual processors. These elements facilitate the use of the Computing Surface in many application areas. An example of a large current system has 400 compute processors.

5.7.2 Caplin Cybernetics HEX

Caplin Cybernetics produce a tranputer based system embedded with VAX/VMS or ULTRIX operating systems. The product, HEX, is made up of processor boards, each with up to eight T800s, interlinked by the HEX highway. The system is interfaced to the VAX via Q-bus, and can have a maximum of 256 processors. This is a multi-user system capable of dynamic configuration and partitioning. Other HEX boards with a variety of functionality may be added to the HEX highway, while connections to other transputers over links can be established via a HXIT board.

5.7.3 Parsys SuperNode

Parsys offer the SuperNode as a supercomputer capable of expansion up to 1024 processors and offering dynamic configuration. It is based upon compute nodes, each with 16 worker transputers; four compute nodes can be connected to a SuperNode switch, which can support up to 72 transputers. By adopting a hierarchical approach, using two levels of switching a SuperNode system can be created with 1024 T800 processors. The system also features the Control Bus, enabling link-independent debugging, synchronization and debugging functions.

5.7.4 Parsytec SuperCluster GigaCube

Parsytec manuafacture the SuperCluster range of systems. These are composed of up to four clusters, each containing 16 T800 processors and an NCU (Network Control Unit 96×96, a communications matrix). The NCU, having enabled full connectivity within the cluster, provides 32 links to allow interconnection between clusters. Additional NCUs in the SuperCluster are used to link the clusters together, their 64 remaining links being available for the interconnection of SuperClusters. Parsytec have also announced a product based upon the T9000, the Parsytec GC; this comprises GigaCubes, each with 64 processors, with a maximum of 16,384 processors and a projected performance of up to 400 GigaFlops.

Appendix: Company contacts

AIKI Parallel Systems
78 Queens Road,
Bristol BS8 1QX
UK
+44 272 272024

Archipel
P.A.E. du Levray,
4, route de Nanfray,
74960 Cran-Gevrier,
Annecy,
France
+33 50 69 20 58

Caplin Cybernetics
Poplar Business Park,
10 Prestons Road,
London E14 9RL
UK
+44 71 538 1716

Digithurst Ltd
Newark Close,
Royston,
Herts SG8 5HL
UK
+44 763 246313

Linkspeed Technologies Ltd
1 Brunswick Street,
Belfast BT2 7GE
UK
+ 44 232 332219

Meiko Ltd
650 Aztec West,
Bristol BS12 4SD
UK
+44 454 618188

Microway (Europe) Ltd
32 High Street,
Kingston Upon Thames,
Surrey KT1 1HL
UK
+44 81 541 5466

Paratech Solutions
163 Brighton Road,
Coulsdon,
Surrey CR5 2NH
UK
+44 81 763 1540

Parsys Ltd
Boundary House,
Boston Road,
London W7 2QE
UK
+44 81 579 8683

Parsytec Computer GmbH
Julicher Str. 338,
D-5100 Aachen,
Germany
+39 241 16 60 00

Perihelion Ltd
33 Bridge Street,
Cambridge CB2 1UW
UK
+44 223 356555

San Diego Levco Corporation
6160 Lusk Boulevard,
Suite C-100,
San Diego,
CA 92121
USA
+1 619 547 2011

Sension
Denton Drive,
Northwich,
Cheshire CW9 7LU
UK
+44 606 44321

References

Ethernet, Distributed packet switching for local computer networks. *Comm. ACM*, **19**(7), (July 1976).

The Ethernet, *A Local Area Network, Data Link Layer and Physical Specifications*, Version 1.0. Xerox, DEC, Intel (September 1980).

Inmos Ltd. (1991) *The Transputer Development and iq Systems Databook*, 2nd edn., pp. 389–404.

Kraus, J.D and Carver, K.R. (1973) *Electromagnetics*. McGraw-Hill, New York, Chapter 13.

Trew, A. and Wilson, G. (1991) *Past, Present, Parallel*. Springer-Verlag, Berlin.

Part III
TRANSPUTER APPLICATIONS

The true test of the transputer approach to building computer systems comes when we seek to answer the questions: Are transputers useful, in practice? Can they be used to construct real applications? Are they any better than equivalent sequential systems? We need to consider these questions both from the end user's point of view, and from the system developer's viewpoint. End users will normally want to know if such systems are faster and more reliable. System developers will ask: How can a problem be decomposed into a set of parallel processes? What software design methodologies are appropriate? How easy is it to parallelize an existing sequential version of an application?

The way in which to answer these questions is to learn from the experience of those who have developed applications using transputers. Therefore, in Part III, we have collected together a number of case studies from a wide range of application areas. In each case, the authors introduce the application area, and explain where and how parallelism can be introduced. After going into more detail on the application, each chapter generally gives some broader guidelines and tips on developing transputer-based software systems for that application area. However, because of the diverse and practical nature of these chapters, the case studies have deliberately not been 'dressed up', nor have the authors been artificially constrained to conform to a standardized style.

The applications

The application studies which are presented in this part are as follows below.

Image processing with transputers (Chapter 6)

This is an intentionally gentle introduction to developing applications software on transputers. It stresses the adoption of a *structured* approach to software design, and warns against the temptation in parallel programming of throwing out all the hard-earned lessons of software engineering which have been learned in the sequential programming world. Image processing is used to illustrate data parallelism, in which a parallel program is obtained by partitioning the data rather than the code.

Porting a Fortran application to a transputer network (Chapter 7)

This is a more extensive (and more specific) study in image processing than the previous chapter. Actually, its focus is not so much image processing itself, but rather the problem of taking an existing sequential Fortran program, and parallelizing it – the so-called 'dusty-deck' problem. An interesting (and more widely applicable) feature of the approach of Morrow and Crookes is the development of an abstract machine as a tool for hiding the parallelism in the final program. Readers who have to parallelize software from other application areas could learn from this particular approach.

Physiological modelling of the human auditory system (Chapter 8)

This case study is a contrast to the previous image processing examples, particularly in its use of MIMD parallelism (or task parallelism). It illustrates the use of parallelism to achieve a more elegant and reliable solution, rather than merely a faster one. Adamson also considers some low level implementation issues which need to be addressed if an optimized solution is required.

Modelling the dispersion of gaseous pollutants in air (Chapter 9)

Simulations which run faster than real time play an important role in crisis management, when in an emergency situation several courses of remedial action can be 'tried out' quickly by simulation, to see what the best course of action might be. In this chapter, Stewart describes his experience with modelling the dispersal of atmospheric pollution. This study raises some important problems which arise in transputer-based systems that are absent in sequential systems; for instance, the problem of load balancing, where the overall system does not behave efficiently because some processors are practically idle while others are working flat out.

Transputer database systems (Chapter 10)

The large amounts of data in a typical database suggest there is an obvious application of parallelism in constructing large, multi-user databases. However, upon closer examination, matters are far from simple, usually because of dependencies in the data. This chapter is not a specific case study, but a broader consideration of the problem of constructing transputer-based databases.

Using transputers in a robot programming and control system (Chapter 11)

Controlling a robot arm with several joints is a computationally intensive task. It is also one which involves rather a lot of low level control and interfacing. The transputer as a standalone device is not ideal as a sole component for this task; but McKeever and Holton come up with a neat solution which involves using a transputer as a coprocessor to a PC. This gives the benefits of using a PC, with all its associated tools for interfacing to the real world; on the other hand, the transputer is used to boost the PC's computing power by acting as a coprocessor. The software design for such a system is an important part of this chapter. Another feature of McKeever and Holton's work is the use of a formal approach to development of their software. Sad to say, such an approach is rather rare in applications work.

Ray traced graphics (Chapter 12)

As a final application area, graphics offers a lot of potential for using parallelism. In this example, McKeag discusses the porting of an existing ray tracing program on to a network of transputers. Although it produces rather nice pictures, this problem is notoriously computationally intensive. McKeag gives a practical account of his experience in parallelizing this graphics application, and concludes that it is possible in this application to obtain useful speed-up with little effort.

More general issues

Taken as a whole, these application chapters raise and illustrate some further important points.

Programming languages

The transputer is not a one-language processor. Although occam is its object code, most of the case studies are presented using languages other than occam: Fortran (Morrow and Crookes), C (Stewart) and Pascal (McKeever and Holton, and McKeag). Occam is most appropriate for very fine grain parallelism, or when using parallelism to give neater, more elegant and reliable solutions rather than faster ones. But existing sequential language expertise does not have to be jettisoned to use transputers.

Porting sequential programs

Parallelizing existing programs is an important skill to master. Two case studies in particular (Morrow and Crookes, and McKeag) deal with this. It is interesting that in both cases, the authors have followed the path of porting the program on to a single transputer first; then, they have developed what amounts to an abstract machine which encapsulates the communication primitives, leaving the remainder of the program as purely sequential. As a general strategy for parallelization, there is much to be said for this approach.

Load balancing

Distribution of the overall workload of a program across a set of processors should ensure that all the processors have roughly even workloads. Depending upon the nature of the application, the distribution strategy may be a simple static one (as in the image processing example of Crookes and Morrow), or it may need dynamic balancing (as discussed by Stewart). This is obviously something which arises only because of the multiprocessor nature of transputer systems, but it can sometimes have a profound effect on the effectiveness of the solution.

Viability

Just because an application can be parallelized, it does not mean that it is worth parallelizing. In many cases, it is not possible to speed things up by a factor of more than four or five, in which case it might be better just to buy (or wait for) a faster single processor, and not bother going through the parallelization process.

The T9000 and its implications

When the new T9000, the associated C104 routing chip, and compilers for occam 3 are all available, people expect that matters will be a little different. The improved speed of the T9000 will open up new application areas (such as real-time image processing, since an image can be transmitted in real time along the new T9000 links); on the other hand, problems which previously required a parallel solution may be solvable on a single T9000! But from the software development point of view, matters will be not so much radically different, just much more convenient. The Virtual Channel Routing capability effectively overcomes the restriction to four links per processor, and should also provide an acceptably efficient implementation of the Virtual Shared Memory programming model, thus making matters simpler for the program developer.

However, if the structured software engineering approach advocated in some of these chapters has been followed, then most of the software developed for T800s will still be suitable for the T9000, with only the bottom communications layer needing modification: it will become much simpler (or may even disappear!). So the outlook for the next generation of transputers is good, not only for those who love to count MIPS, but also for those who are concerned with higher level approaches to system development. It is rare for both of these outlooks to improve at the same time. Thus, the work presented in these chapters, although targeted mainly towards the T800 series, will not be redundant. The design issues will carry over into the next generation, and the experience which the reader can benefit from will not be wasted.

R.M. McKeag and D. Crookes

Chapter 6

Image Processing with Transputers

D. Crookes and P. J. Morrow

6.1 Introduction

This chapter aims to show how transputers can be used to perform parallel image processing, and to enable the reader to design useful software systems for low-level image processing on transputers. By way of introduction, an overview of some applications is given, followed by a simple programming view of image processing. Then, a practical approach to writing *parallel* image processing software is given. The program code which is developed in the course of the chapter is presented in occam 2. The approach we present is straightforward and methodical, and has been tried and tested. By the end of the chapter, the reader should be in a position to develop basic transputer-based low-level image processing systems.

6.1.1 Commercial applications of image processing

The number of potential industrial applications for computer image processing is large. The range is increasing as the hardware technology becomes cheaper and faster, and as the software becomes more effective. The following areas represent some of the more significant applications:

Industrial inspection

Inspecting products coming off a production line is a key part of quality control in any manufacturing process. The products may be biscuits which need to be inspected for proper shape, or they may be printed circuit boards which need to be inspected for breaks in solder lines, or unintended junctions where two lines of solder have touched. Depending upon the type of production faults which are known to be likely, the inspection may be very detailed and intensive, or it may be cursory. Some inspection processes will sample only a selection of products, if the requirement is merely to identify a fault somewhere in the production line. In other cases, it may be necessary to inspect every single item, with the aim of ensuring that no faulty products are allowed to be passed on to the customer. Clearly, then, the amount of time, effort and skill required by the inspection process depends very much upon the product in question and on the purpose of the inspection process.

Because repetitive inspection by humans is a boring and labour-intensive job, there can be advantages in automating the process by using computer image processing. A computer can work round the clock, seven days a week, without complaining. More importantly, a computer does not lose concentration or go to sleep during the night shift, and let faulty products go undetected. For reasons such as these, automated inspection facilities have been successfully installed for the inspection of products such as PCBs, textiles, bricks, biscuits, metal castings, micro-electronic components, and many others.

Target tracking

Following a moving object across a scene is a subject of great interest to many people, even if the object being tracked is a target for nothing more destructive than a speeding fine. There are several obvious military applications, such as guiding a missile more accurately by using vision rather than by crude heat-seeking. But there are also many civilian applications, such as non-clandestine satellite surveillance, air traffic control and robotic applications where the robot is interacting with objects which are moving relative to the camera.

This type of application generally requires a bit more intelligence than, say, industrial inspection, since a certain amount of knowledge is required to distinguish the object from the background. Distinguishing and recognizing different objects against a real life background is a difficult task.

Remote vehicle guidance

There are many working environments in which it would be very dangerous or even impossible for humans to operate. Monitoring and maintaining the inside of a nuclear reactor, or prospecting in murky waters at the bottom of an ocean, or operating in the depths of outer space, are applications which call for a remotely-controlled vehicle with some vision capability.

This type of application has its own particular problems when it comes to automation. The fact that the camera is itself moving does not help. Neither does the fact that lighting conditions (a crucial factor in image processing) cannot usually be controlled.

Image transmission and storage

An ability to transmit moving pictures in real time along digital networks would open up a large number of applications. There are already initial versions of video telephones and video conferencing systems. The overriding problem with this type of application is, of course, the vast amount of data which needs to be transmitted. To send a moving monochrome picture (at 25 frames a second) with a reasonable resolution of, say, 256 × 256, would appear to require a bandwidth of about 12 Mbits/second. Given that a telephone line can handle a transfer rate of the order of 0.01 Mbit/second, the problem becomes obvious. The solution is to do a large amount of image compression, by encoding each picture (or the changes from the previous one) in some compressed format before transmission, and then reconstructing the original image at the receiving end. This involves a large amount of image processing in a very short time; so the computational requirements can be enormous. With the prospect of High Definition TV (HDTV), where the image resolution could be in the region of 1250 × 1250, the computational requirements are easily an order of magnitude greater.

There are also many applications which require the storage of large numbers of images. People are realizing the potential for image databases in, for instance, museums and art galleries. Hospitals are particularly heavy users of stored images in the form of X-rays, which need to be stored and retrieved. Even with high capacity storage media such as videodisk and CD-ROM, image compression still offers considerable economic advantages. Admittedly, these applications may not have to operate in real time, at 25 frames/second; but there is still a time constraint in image retrieval and decompression.

Medical image processing

There are many aspects of medical practice which are benefiting from the use of computer image processing techniques, even if the 'images' are not what could be seen by the naked eye. EMR scanners and ultrasound scanners produce non-visual signals which, if processed appropriately, can be displayed as a image.

One aspect which is particularly useful in medical applications is to automate the measurement of objects detected in an image. This automatic measurement can be a real time saver, for instance, in experiments in which the behaviour of cells is being observed over a period of time. Using appropriate techniques, the computer can calculate areas and rates of change which would otherwise be very tedious to measure manually.

6.2 Image processing tasks and techniques

This section identifies some of the tasks which are commonly performed on images, and outlines some of the basic algorithms and methods used to perform these tasks.

6.2.1 Common image processing tasks

Noise reduction

A captured image is rarely an exact representation of the viewed scene: a certain amount of noise in the image is inevitable. This can show itself as speckle, or as grain, or as blurring at region boundaries. Usually, the first stage in processing an image is to perform some noise reduction.

Simple noise reduction techniques rely on inspecting a small neighbourhood (say, 3×3) around a pixel, and trying to work out whether the centre pixel is a genuine pixel value, or one affected by noise – and if it is a noise-affected pixel, what adjustment to its value is necessary to restore it. One simple approach is merely to replace every pixel with the average of its 3×3 neighbourhood. This removes a piece of random noise in an otherwise smooth region, and will leave a genuine smooth region unaltered. But unfortunately, the averaging operation will also blur edges in the image. A better method in many cases is to choose the *median* value in the neighbourhood instead of the average (for instance, by sorting the neighbourhood, and choosing the middle value).

More sophisticated noise reduction techniques try to model the cause of the noise, and therefore should be in a better position to differentiate between 'true' noise and genuine image detail. For instance, it often happens that the noise has a characteristic frequency, which can be removed from the image. This is achieved by first finding all the different frequencies in the image, using the Fourier Transform; then, masking out the characteristic frequency of the noise; and finally converting back the filtered frequency image into the original spatial domain. A standard algorithm for this process is the Fast Fourier Transform (FFT). FFT is not a neighbourhood-based algorithm: each pixel of the result depends on the entire original image.

Contrast control

When lighting conditions are not ideal, it can happen that important detail in an image is not as obvious to the human eye as it should be, because of poor contrast between shades. Simple contrast improvement involves lightening the lighter parts and darkening the darker parts. More sophisticated approaches may improve the contrast only in certain regions of the image, or by different amounts in different regions depending upon the nature of each region. Also, one may need to take account of the fact that the human eye is not equally sensitive to all shade differences: the eye is more sensitive to shade differences in a dark region than to the same differences in a light region.

Edge detection

When it is necessary to analyse the content of an image, the best starting point is often to find the edges. An edge shows up as a relatively sharp change in brightness within a small region, and so can be detected by a neighbourhood-based operation. One fairly general-purpose method is the Sobel edge detector. Sobel effectively differentiates each 3×3 neighbourhood in the image horizontally and vertically, and identifies a pixel as an edge pixel if its combined horizontal and vertical gradients are above a suitable threshold.

Note that the idea of *neighbourhood* processing is emerging as a fundamental operation, in which a pixel of the processed image is found by placing a small 'window' (often 3×3) over the pixel in the original image, and doing a calculation based solely on the values seen in the window. Because of the very localized nature of this operation, different parts of the result image can be calculated simultaneously, in parallel. This is another attractive feature of the neighbourhood approach.

Image compression

When compressing a (still) image, a key property of an algorithm is whether or not it enables the original image to be reconstructed with no loss of information (so-called lossless compression). In general, the higher the compression ratio required, the more information has to be discarded during the compression process. The trade-off depends upon the application.

The most common technique for image compression is based on the Discrete Cosine Transform (DCT). An image is first broken down into small blocks (typically 8×8 pixels), and each block is compressed separately. Each block is converted into its frequency components (as in the FFT), in which the high frequencies correspond to the very fine detail. By discarding frequencies from the high end, we throw away some of the very fine detail in the block (some of which may not even be visible to the eye). The more frequencies are discarded, the greater the compression, and the less accurate will be the reconstructed image. The retained frequency coefficients are then encoded using a technique known as quantization. For a 'typical' monochrome image of, say, a landscape scene, a compression ratio of up to 10:1 can often be achieved with only a small amount of visible distortion.

In the DCT approach, the processing required is again localized, because the image is compressed as a set of independent little blocks. This makes it amenable to parallel implementation, in which different parts of the image can be compressed at the same time.

Another promising compression technique is based on *fractals*. Fractals are mathematical functions which can be used to generate graphical patterns and textures, some of which happen to look like naturally-occurring patterns and surfaces. Fractals can be used for compression by modelling the regions and features of an image with fractals, enabling the original features (or something like them) to be regenerated at a later date. Claims of very high compression ratios have been made for certain applications. Unfortunately, from a user's point of view, the algorithms involved in this process are not yet in the public domain, and there is still much work to be done in evaluating the performance and suitability of fractal techniques. However, results to date would indicate that general purpose fractal techniques will be able to offer compression ratios several times greater than the DCT approach, with acceptable computational requirements. Fractal compression could well make current techniques, such as DCT, redundant in the not too distant future.

Image segmentation

To begin to recognize objects in an image, it is often necessary to detect *regions*. A region, or segment, is a set of (connected) pixels which share a common property. The property might be having a similar texture, or a similar distribution of grey levels.

There are several approaches to finding regions, including the following:

- Choose a 'seed' point somewhere in a region, and grow the region outwards, rather like paint spreading until it meets a boundary.
- Subdivide the image into four quadrants; test each quadrant to see if all pixels in it share the desired common property (and hence form part of a region); if not, then split or subdivide the quadrant, and repeat the process recursively. By the end of this phase, the whole image will have been decomposed into a large number of small squares of different sizes. The second phase then involves trying to merge adjacent squares if they are part of the same region. This ends up with a smaller number of irregularly-shaped regions (with a blocked appearance). This approach is known as 'Split-and-Merge'.

Pattern recognition

Once we enter the realm of trying to recognize *objects* in an image, we are trying to match the perceived features in the image (lines, regions, etc.) with a description of objects which might be expected to be present. This matching process involves testing various hypotheses, rejecting some, following up others, backtracking when a hypothesis is eventually rejected, and other complicated control strategies. At this level, the data includes not only the features in the image, but the object descriptions from a database of possible objects

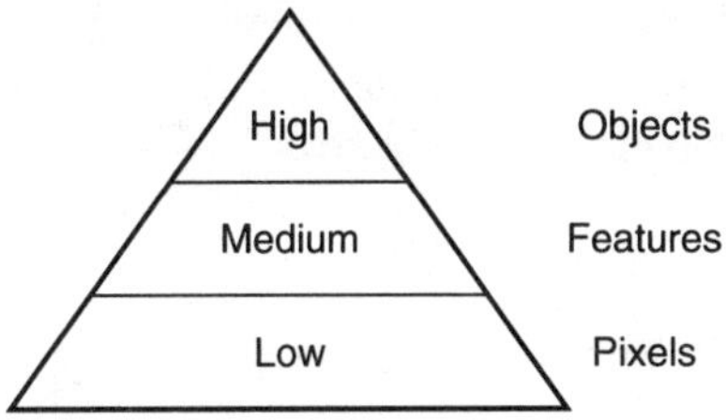

Figure 6.1 Relationship between levels represented as a pyramid.

(that is, the hypotheses which have to be tested). Because of the trial and error nature of the matching process, it is not so obvious how to apply parallelism at this level of image understanding.

6.2.2 The image processing hierarchy

The range of typical operations presented above can be classified into three broad categories, or *levels*, by noting the type of data which the operation consumes and generates. The three categories are generally taken to be as follows:

- *Low level-operations* These take a pixel image, and generate another pixel image. Examples include contrast adjustment and edge detection.
- *Medium-level operations* These take a pixel image, and generate a set of features, such as a set of lines or regions.
- *High-level operations* These take a set of features, and generate a set of objects.

Because the amount of raw data decreases as we move up the hierarchy of levels (from a large number of pixels at the low level to a small number of objects at the high level), the relationship between the levels is often presented as a pyramid, as shown in Figure 6.1.

6.3 Transputer-based parallel image processing

The justification for looking to parallel processing in image processing systems can be summed up in one word: *speed*. Because of the large amount of data being processed, and the (usually) limited time available, ordinary sequential computers either cannot do the job fast enough, or are too expensive to be cost-effective.

One advantage which transputer-based parallel solutions offer over other parallel computers is **scalability**. The amount of processing power can be selected to match the real-time requirements of the problem. It will also be important, therefore, to make sure that in programming transputer systems, the software is also scalable.

6.3.1 Parallelism in image processing

There are two common approaches to applying parallelism to applications such as image processing: **task** parallelism, and **data** parallelism.

Task parallelism

Since a complete image processing system usually requires a sequence of different tasks to be performed on an image, it can sometimes be possible to have these tasks running concurrently on different processors. For instance, if a video sequence is being processed, then once one task has processed its first image, it can pass the intermediate result on to the next task, and immediately take in the next image in the sequence and start processing that. Even when processing a single image, the system can sometimes be pipelined so that tasks further down the line can receive and begin processing part of the image before previous tasks have completed processing of the rest of the image. This MIMD parallelism is sometimes called **task parallelism**. It suffers, however, from two main disadvantages. First, as with all pipelined systems, the rate of throughput is dictated by the speed of the *slowest* task in the pipeline. Second, the software is not *scalable*, since it is usually impossible to decompose a problem into an *arbitrary* number of concurrent tasks.

Data parallelism

A more obvious (and generally more effective) way to parallelize image processing operations is to partition an image into sections, distribute these sections across a network of transputers, and have all transputers carry out the same task on their own section of the image. This is an application of data (SIMD) parallelism. For low level image processing, where pixel-level and neighbourhood operations predominate, we have already seen that the operations are very localized, and so the transputers can usually operate more or less independently, and in parallel.

There are two main approaches to implementing data parallelism on transputer networks: a static approach, and a dynamic approach.

Static data distribution

Here, image data is resident on the transputer network. An image is partitioned into an appropriate number of segments, and distributed one segment per transputer, as illustrated in Figure 6.2 in the case of four transputers. All transputers maintain and process their own (static) segment of the image simultaneously.

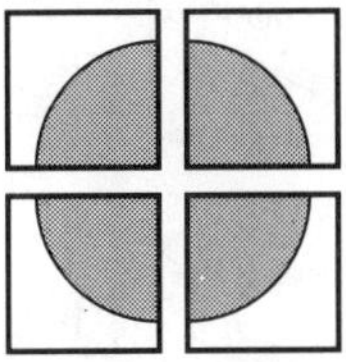

Figure 6.2 Static data distribution.

This simple approach is very satisfactory in many cases. Processing can often proceed independently on each transputer, with very little communication. A slight problem arises when performing a neighbourhood operation at the edge of a segment – say, the right-hand edge of the top left segment in Figure 6.2. For a pixel on this edge, the neighbourhood appears to straddle two transputers (top left and right). To overcome this problem efficiently, the usual technique is to extend each segment beyond the bare quadrants, and hold overlapping borders on each transputer. Thus, the boundary pixels for each segment are duplicated on neighbouring transputers. If the neighbourhoods being used are never more than 3×3, then a one pixel-wide border extension is all that is necessary.

However, problems can arise with this static approach when the *information* content of the image (for example, the edge information) is unevenly distributed throughout the image. More generally, when working in parallel with features of an image (such as lines or regions) rather than just pixels, the problem of **load balancing** is bound to raise its head. This is because, although pixels are evenly distributed in an image, the information content (and hence the workload) is usually unevenly distributed. Attempting to distribute the image in a way which balances the workload is called load balancing. The static distribution strategy outlined above is susceptible to load balancing problems, and for this reason it may be worth considering the alternative of a more dynamic approach.

Dynamic data distribution

The method of processing an image by **farming** involves holding the image on a controlling processor, dividing it up into small packets, and 'farming out' the processing of these work packets to a pool of worker transputers. The worker processors return result packets, and wait for another work packet to process, as illustrated in Figure 6.3.

If a worker receives a work packet which requires very little computation, it will quickly be available to receive another; a 'heavy' work packet, however, will keep a processor busy for longer, and so it will not process as many packets. This flexibility enables better load balancing to be achieved when the workload is unevenly distributed throughout the image.

Although providing better load balancing, the farming approach does have its drawbacks. The main problem is the amount of communication involved.

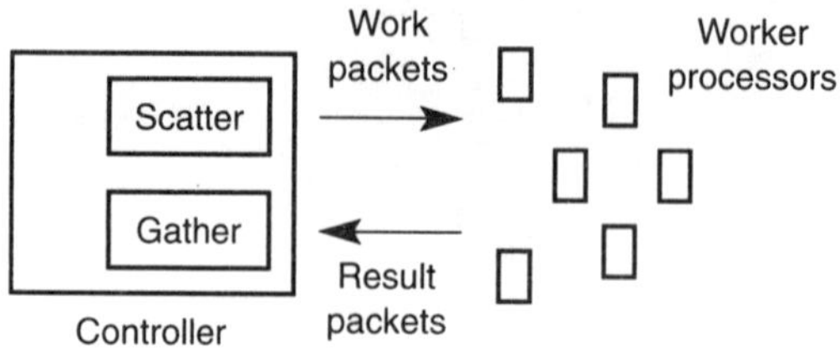

Figure 6.3 Dynamic data distribution.

If the computational effort involved in the entire operation is relatively small, then the overheads in farming out the work packets and receiving the result packets can considerably outweigh the saving from improved load balancing. The choice between a static and a dynamic distribution depends upon the amount of computation involved in processing a work packet, as well as the unevenness of the workload.

Whichever way of partitioning the image data is chosen, it is inevitable that some operations will prove difficult or impossible to do in parallel. Contour tracing, in which we start at a given point on an edge and follow an outline which could potentially traverse the whole image, is one example; or, region growing in which we start from a given seed point and grow the region out to the region's boundaries (like spreading paint) is another. In general, where we are trying to find features whose pixels straddle the boundaries of a distributed segment, we can expect problems. Once the features have been found, however, it is often possible to avoid some further parallelization problems by gathering the set of features together, and then broadcasting this set dynamically to each transputer. Thus, each transputer can have its own copy of all the features, even if it has to process only its own section of the feature set. Since by this stage the data will have been reduced in its amount, this replication strategy is usually feasible within typical memory constraints.

6.3.2 Transputer architectures for image processing

A complete transputer-based image processing architecture has several components; a typical host-independent configuration might be as shown in Figure 6.4.

The computing components are sometimes provided as separate boards which plug into the host computer's bus, though the motherboard + TRAM format is becoming more standard. A typical set-up will use a PC as a host front-end for the overall system. The following are some practical details about current products in this area:

- *The **frame grabber**, for image capture* Examples include the Sequin (monochrome) and Mosaic (colour) frame grabber boards from Quintek, which can capture at a resolution of 512 × 512 pixels. There is also a

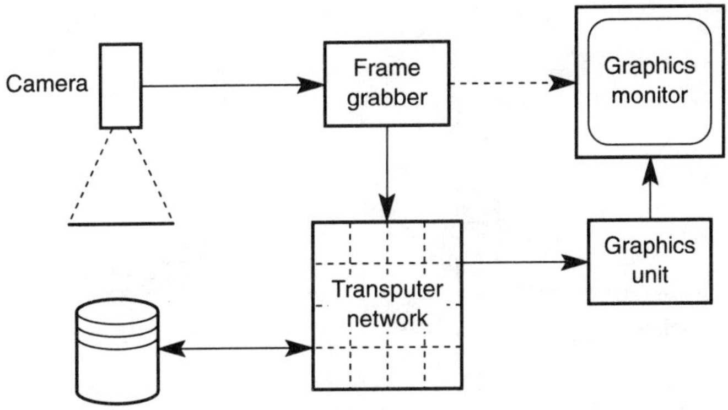

Figure 6.4 Typical host configuration for an image processing architecture.

board from Digithurst, the MicroEye TC (colour) or TM (monochrome), which has a higher resolution. In TRAM format, examples are the TTGF from Transtech, or the Inmos IMS B429 Video Image Processing TRAM. The latter has powerful on-board special-purpose image processing chips for very fast initial preprocessing of the captured image before it is passed on to the rest of the system for further analysis.

- *The **graphics unit** for image display* There are several TRAMs which offer monochrome and colour display capabilities at resolutions ranging from 512 × 512 to 1 K × 1 K and beyond. A common and convenient arrangement, combining the frame grabber with the image display on the same board, is provided by the Digithurst frame grabber board, amongst others.
- *The **disk** for image storage* A simple and cost-effective arrangement is to use the disk on the host machine (for example, a PC) for image storage. This tends to be slow for data transfer, but can be speeded up if a 16-bit DMA transfer capability is available (as provided by some motherboards). A more sophisticated approach is to use a large SCSI disk interfaced directly to the transputer network by a SCSI TRAM. This invariably requires extra software development of a rather technical nature.

Transputer network configuration

In a way, the very flexibility of transputer network configurations is itself a problem, in that it makes the software less likely to be portable across other transputer networks. In practice, it is better to use a fairly standard configuration which is, nevertheless, scalable. From our own experience, we would recommend the configuration shown in Figure 6.5.

This configuration enables images to be transferred between the network and the image Input/Output devices using all four links. To scale the processing

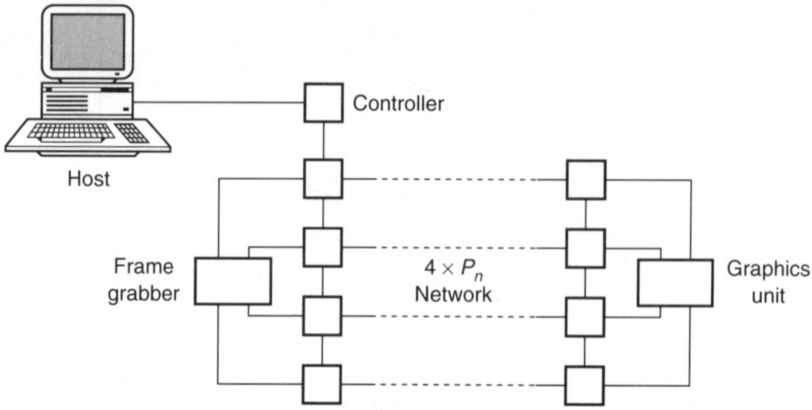

Figure 6.5 Recommended transputer network configuration.

power, extra columns of four transputers can be added. (It is usual, but not essential, for P_n to be a power of 2.) If the hardware forces the frame grabber and the graphics unit to reside on the same processor (as mentioned above), then the graphics unit on the right of the diagram can be dropped from the design, and the frame grabber processor becomes a joint frame grabber/graphics unit.

6.4 Software design approach

It is possible to achieve good results in transputer-based low level image processing by following a simple and standard model for developing software. We recommend the following guidelines, which have proved widely acceptable:

- Use data parallelism, rather than task parallelism.
- Use static data distribution for images, rather than a dynamic farming approach.
- Configure the transputer network as a $4 \times P_n$ grid of nodes, as in Figure 6.5.
- Assume the image display unit is separate from the frame grabber.
- Use the host's disk for image files.
- For every image, hold an overlapping border around each segment, whose width is determined by a compile-time constant called max.border. The value of max.border will depend upon the largest neighbourhood that will be permitted.

In what follows, we have used occam 2 as the programming language when presenting code. However, the principles will be the same even if a different programming language were to be used for coding. It is also assumed that the images being processed are *monochrome* rather than colour.

6.4.1 Process structure

The structure of the overall program matches the architecture illustrated previously, and comprises:

- a $4 \times P_n$ array of node processes, which do the actual image processing. All nodes have the same code, though each node process will declare and process only its own segment of every image. A node process knows its own identity, so it can calculate where it is in the network;
- a controller process, which mainly acts as a server to the $4 \times P_n$ grid of nodes;
- a graphics process for image display;
- a frame grabber process. Depending upon the hardware used, this may have to be multiplexed with the graphics process.

In occam, this framework can be coded in a straightforward manner. First, the following processes are defined:

```
PROC node (...)
PROC controller (...)
PROC graphics (...)
PROC framegrabber (...)
```

and then instantiated and placed on the network in the configuration section. Only one instance of the latter three processes are required, but a $4 \times P_n$ array of thc *node* proccss will bc nceded.

During execution, the network of nodes acts as the master, and the other three processes (controller, graphics and frame grabber) act as servers to the node process array. When the node array requires a service from one of these supporting processes (usually for input or output), it will first send a command indicating the service required. The data transfer will then take place (Figure 6.6).

Since the three supporting processes are command driven, their structure and content will be determined by the node process array which drives them. The content of a node process is therefore the most crucial aspect, and is the main focus of what follows.

6.4.2 The node process

The node process is where the actual image processing is done. Each node processes a fixed area of all images. Before considering the details of coding image processing operations on the node, we must first decide on a standard image representation.

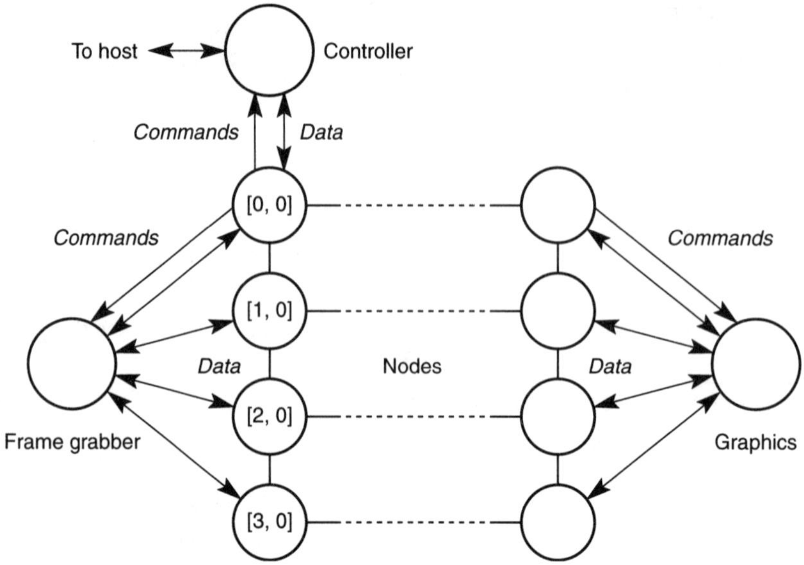

Figure 6.6 Data transfer.

Image representation

An $N \times M$ image will be distributed over a $4 \times P_n$ array of nodes by partitioning it into equal segments of size `num.rows × num.cols`, where:

```
num.rows = (N / 4)
num.cols = (M / Pn)
```

As we have seen, in addition to the actual segment, an overlapping border must also be accommodated. An image segment thus has the form shown in Figure 6.7, where:

```
max.rows = (2 * max.border) + num.rows
max.cols = (2 * max.border) + num.cols
```

One further convention which is often adopted is always to represent pixels as real (floating point) values, rather than integers. One reason for this is that on a T800 transputer, real arithmetic is usually faster than integer arithmetic. Also, when writing standard procedures, it means that we don't have to write one version for integer images and another for real images. Problems with this convention are that we cannot save storage by using byte images, and there can also be problems when comparing two floating point pixel values for equality. But in general, the convenience and speed advantages outweigh these disadvantages.

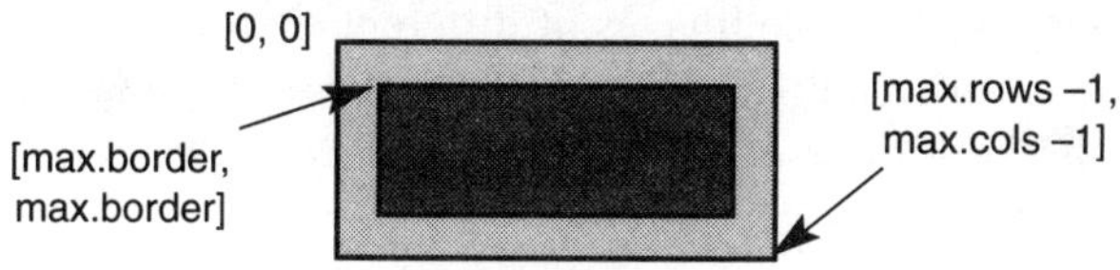

Figure 6.7 Image segment form.

We are now in a position to give the occam declaration which should be included in each node process to represent an image, say Image1:

```
[max.rows][max.cols] REAL32  Image1:
```

Non-image data should be replicated across all the nodes. For instance, if a threshold value is to be used, it should be declared and assigned in each node process. Although there is apparent redundancy in this replicated processing, it is faster than having just one node process performing the operation and communicating the results to the other nodes. Likewise, other arrays which are not going to be processed in parallel should be replicated, and not distributed.

Simple parallel image processing

Given an image declared and distributed in keeping with the guidelines above, processing this image in parallel merely involves each node process performing the required operation on its own image segments. We will now look at the coding of a range of typical image processing operations, beginning with simple point-to-point operations, then neighbourhood operations, and finally an approach to coding global operations.

Point to-point operations Consider first the case of point-to-point operations (such as image subtraction). The general case in which an image I1 is to be calculated from two operand images I2 and I3 using a pixel operator <op> could be written as the high level operation:

```
I1 := I2 <op> I3
```

To implement this in parallel, each node process would merely calculate its own segment of the image I1 as follows:

```
SEQ row = max.border FOR num.rows
  SEQ col = max.border FOR num.cols
    I1[row][col] := I2[row][col] <op>
                                      I3[row][col]
```

Sometimes we may wish to use images of different sizes in the same program. To avoid having to write a different procedure for each image size, the loop bounds should be kept general by writing them in terms of the size of the images being processed. The occam function:

```
SIZE  array
```

can be used for this purpose. So, for instance, a point-to-point image subtraction procedure could be made available in a more general form by first precalculating `n.rows` and `n.cols`:

```
PROC image.image.subtract ([][]REAL32 I1, I2,
                                               I3)
  INT n.rows, n.cols:
  SEQ
    n.rows := (SIZE I1) - (2*max.border)
    n.cols := (SIZE I1[0]) - (2*max.border)
    SEQ row = max.border FOR n.rows
      SEQ col = max.border FOR n.cols
        I1[row][col] := I2[row][col] -
                                 I3[row][col]
:
```

A set of similar procedures could be written for other point-to-point operations (see later for a suggested useful range of such procedures).

Neighbourhood operations Neighbourhood operations are slightly more involved because of the need to make use of the overlapping border regions. However, in the simplest kind of neighbourhood operation, the general processing structure will be rather similar to the above; consider, for instance, the generation of a result image `I2` based on a 3×3 neighbourhood operation on an image `I1`:

```
SEQ row = max.border FOR n.rows
  SEQ col = max.border FOR n.cols
    I2[row][col] := operation applied to the
                              3x3 neighbourhood
                    I1[row-1..row+1][col-
                                   1..col+1]
```

Unfortunately, there are some complications with this simple formulation which must be dealt with. The most important is to note that, since this coding will reference values in the border region of `I1`, then it must be guaranteed that all these border regions are in fact completely up to date in their contents before the neighbourhood operation is performed. As things stand, this might well not be the case: note, for instance, that the above coding of a point-to-point (subtraction) operation will not update the border regions of its result image `I1`, and therefore will leave `I1`'s borders in an invalid state. This is not important if `I1` is to be used subsequently for another point-to-point operation; but it is crucial if `I1` was the operand of a *neighbourhood* operation such as that above. A simple solution is to guarantee, immediately prior to any neighbourhood operation, that the borders of the operand image are properly updated. This can be done by making use of a standard runtime support routine (yet to be written!) which updates all border regions of an image segment by exchanging appropriate borders with the four neighbouring processes of the node:

```
swap.borders (image, border size)
```

The supplied border size is determined by the size of the neighbourhood about to be used. This could in some cases be less than `max.border`, if neighbourhoods of different sizes are being used at various points in the program. To be absolutely sure that borders are valid, this procedure could be called immediately prior to every neighbourhood operation. However, there could be some wasted effort in doing this, if the borders of an image happened in fact to be perfectly up to date (perhaps because the operand image had already been used as an operand in a previous neighbourhood operation).

Image boundary effects The approach so far has masked one other problem, namely what happens during border swapping at the extreme outer edges of the image (which are resident on the transputers on the outside rim of the transputer grid)? These outer borders do not overlap with any adjacent segments, so what values should these border regions be given? What we have in fact done so far is to add on a little strip the whole way round the image, without defining the contents of this strip. There are several possibilities for defining this strip during border swapping:

- set it all to black (grey level 0);
- set it all to white (grey level 255);
- extend the edges of the image out, and fill the strip by duplicating the outer edge;
- wrap the image round, so that the left hand strip is set to the right hand border of the image, and so on. In other words, the image is made to look like a **torus**.

The first two approaches both give a discontinuity in the image which might be unwelcome. The fourth possibility involves a communication pattern which is rather tricky to implement when the transputer grid links are not themselves connected as a torus. (Since this would use up all four links of each transputer, a torus is difficult to arrange if you want to get any information in or out of the grid!) In practice, we recommend the third approach, of border replication. If you really want to retain flexibility, you should add another parameter to the swap.borders routine to indicate which option you want.

Now let's extend the above basic structure by including the use of an explicit window, or **template**, which represents the neighbourhood and which

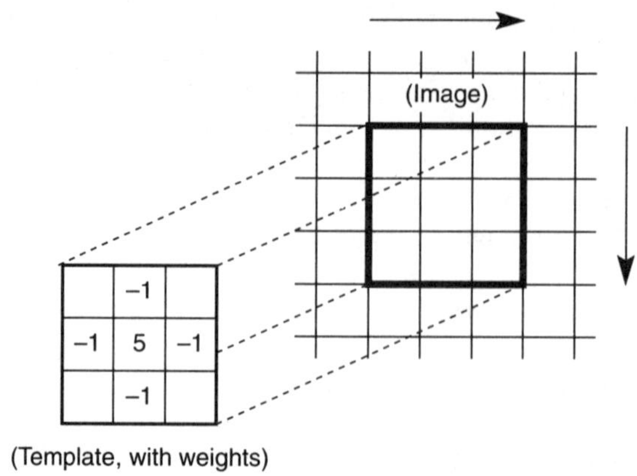

Figure 6.8 An explicit window with an additional value or weight corresponding to each image pixel in the neighbourhood.

moves over the entire image. Many useful effects can be achieved by having an explicit window which has an additional value or **weight** corresponding to each image pixel in the neighbourhood, as illustrated in Figure 6.8.

The most common use of a template is to perform a **convolution** of the template with the image. In convolution, the result pixel for a given window position is calculated by multiplying each pixel in the image neighbourhood by the corresponding weight in the template, and then summing these products. (We are assuming that the result pixel goes in the position corresponding to the *middle* of the window. This is the commonest usage, though sometimes it can be useful to vary this position.)

This simple form of convolution can be extended in several ways:

- The template size need not always be 3×3.
- Not all weights in the template need to be defined. An undefined weight will cause the corresponding image pixel to be excluded from the neighbourhood. This is quite common, and means we can define non-rectangular neighbourhoods. For instance, in the diagram above, the template defines a cross-shaped neighbourhood.
- The multiply-and-add sequence which defines a convolution is not the only kind of neighbourhood operation. Other useful operations are:
  - *absolute convolution* (positive value of a convolution)
  - *multiplicative maximum* (multiply weights and pixels, and find the maximum, instead of summing)
  - *multiplicative minimum* (multiply weights and pixels, and find the minimum)
  - *additive maximum* (add the weigh ts and the pixels, and find the maximum)

	Row	Col	Weight	
Length	5	3	3	Window size
	–1	0	–1	
	0	–1	–1	
	0	0	5	
	0	1	–1	
	1	0	–1	

Figure 6.9 The template in Figure 6.8 held as a structure.

- *additive minimum* (add the weights and the pixels, and find the minimum)
- *median* (find the median value of the pixels in the window; weights are not used).

To represent a template, we need to store not only the weights, but which template positions are defined. A simple and reasonably efficient general way of doing this is to hold a list of the defined template positions (as row and column offsets from the centre), and their corresponding weights. The length of this list depends upon the number of defined positions, and needs to be stored as part of the template, as does the size of the rectangular window. For instance, the template in Figure 6.8 could be held as the structure shown in Figure 6.9, in which the first row holds the latter three pieces of housekeeping information.

A template like this could be declared in occam as a 2D array, whose size depends upon the number of potential weight positions in the window:

```
[10][3] INT  Template1:
```

Given this representation, a more general procedure for performing a convolution on an image `In` with a template `T` giving a result image `Result` could be written as:

```
PROC convolution ([][]REAL32 In, Result,
                                      [][3]INT T)
  INT n.rows, n.cols, border:
  SEQ
    border := (max(T[0][1], T[0][2]) / 2)
    swap.borders(In, border)
    n.rows := (SIZE In) - (2*max.border)
    n.cols := (SIZE In[0]) - (2*max.border)
    REAL32 sum:
    SEQ row = max.border FOR n.rows
```

```
        SEQ col = max.border FOR n.cols
          sum := 0.0 (REAL32)
          SEQ i = 1 FOR T[0][0]
            sum := sum + ((REAL32 ROUND T[i][2]) *
                       In[row+T[i][0]] [col+T[i][1]])
          Result[row][col] := sum
  :
```

(This coding assumes that a simple function max has been defined which returns the maximum of two integers.) A set of similar procedures could be written for the other neighbourhood operations mentioned above.

To avoid redundant effort within `swap.borders` in these neighbourhood procedures (when the borders are already valid), it would be necessary to maintain and pass a boolean flag (called, say, `border.valid`) for each image, indicating the current validity of that image's borders. This would be passed to `swap.borders`, which would test the flag, and only update the borders if the flag was `FALSE`, setting it back to `TRUE` on exit from `swap.borders`. It would also be necessary, whenever a new image is calculated, to set this flag to `FALSE` as the final stage in the calculation. Although this modification can lead to greater efficiency, occam 2's lack of a record-like data structuring facility makes the resulting procedural interfaces a little less elegant.

Global operations Unlike the operations we have considered so far, some quite common operations are global in nature, in that the operation cannot be carried out with no communication among the processors.

Suppose, for instance, we need to calculate the average grey level for a whole image, and so wish to add up all pixel values. It is simple enough for each processor to sum its own segment, giving one value on each processor. But these distributed partial sums have then somehow to be accumulated. A similar problem arises when we want to evaluate the maximum or minimum pixel value in an image: again, it is simple for each processor to evaluate its local result, but the distributed partial results must be compared to enable each processor to know the final, global result. Rather than treat each of these as a special case, each requiring its own standard runtime support routine, it is neater to define a single standard routine (still to be written) which merely gathers together a distributed set of values, one per processor, and ensures that every processor has a copy of every value:

```
PROC gather.values (VAL REAL32 local,
                              []REAL32 all.locals)
```

The size of `all.locals` should be the number of transputers in the grid. All the necessary communication is hidden within this procedure. Assuming this routine, it is not difficult to code global operations such as image sum, maximum or minimum:

```
[4*Pn]REAL32 local.values:
REAL32 local.value
-- calculate local partial result,
-- in local.value
...
gather.values (local.value, local.values)
-- calculate final result from the list of all
-- local.values
...
```

Slightly more complex operations can be handled in a similar way. For instance, finding the **histogram** of an image is commonly used to discover the spread of grey levels in an image. Again, this operation can be coded as a local operation (finding the histogram of the local image segment), followed by a 'gather' operation, followed by replicated calculation of the global final result by adding together all the partial histograms. This requires a slightly different gather routine, since it gathers arrays rather than single values:

```
PROC gather.arrays (VAL []REAL32 local,
                                [][]REAL32 all.locals)
```

6.5 A layered software model

In the section above, we have advocated the development of quite a number of utility procedures which should prove rather useful when putting together actual programs for image processing tasks. There have been procedures for point-to-point operations, for swapping borders, for neighbourhood operations, for gathering values, and others. Some of these involve some awareness of communication (for example, `gather.values`), while others do not. For developing programs, it is good practice where at all possible to shield the programmer from lower level details (such as communication). So ideally, a programmer should be presented with a set of image processing routines (like `convolution` and `image.image.subtract`), but not necessarily the communication routines (like `swap.borders`). Therefore, the software running on the node processors should be structured as a set of layers, as illustrated in Figure 6.10.

In practice, because the set of standard image processing routines cannot possibly be complete, the programmer will have to add others, and so will need access to the communication routines. Indeed, it is likely that new communication routines will have to be added for certain algorithms. Nevertheless, this layered model gives a very useful framework which should be retained even though it is added to.

The *communications* layer should include at least the following routines:

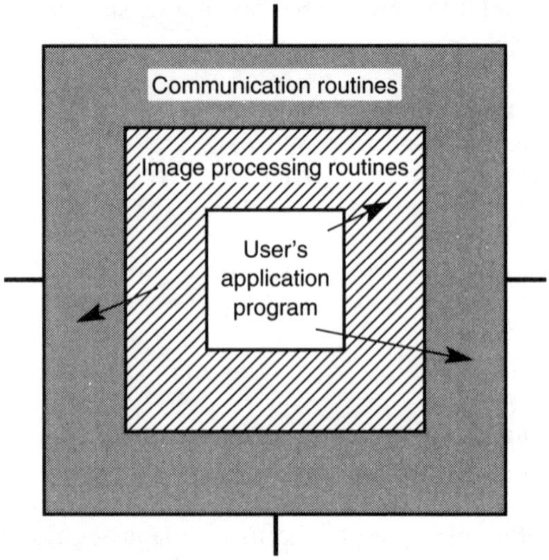

Figure 6.10 Software structured as a set of layers.

- For image input and output:
  - read an image from a file on the host's disk (via the controller),
  - grab an image from the frame grabber process,
  - put the video camera into live mode,
  - display an image on the graphics display (say whether the image should be scaled or truncated to be in the range 0..255),
  - write an image to a file on the host's disk (via the controller).
- For data communication between nodes:
  - `swap.borders`
  - `gather.values`
  - `gather.arrays`
- For standard communication with the host (via the controller):
  - read scalar values (integers, reals, characters or strings) from the keyboard,
  - read scalar values from a file on the host's disk,
  - write scalar values to the host's screen,
  - write scalar values to a file on the host's disk.

The *image processing* layer should include at least the following routines:

- For point-to-point operations between two images:

```
– image.image.add          (I1 + I2)
– image.image.subtract     (I1 - I2)
– image.image.multiply     (I1 * I2)
– image.image.divide       (I1 / I2)
```

- `image.image.equal` `(I1 = I2)`
- `image.image.not.equal` `(I1 <> I2)`
- `image.image.greater` `(I1 > I2)`
- `image.image.greater.equal` `(I1 >= I2)`

  (less than or equal, and less than, can be obtained from the latter two).

Boolean values should be represented using 1.0 for *true* and 0.0 for *false*

- For point-to-point operations between an image and a scalar (REAL32) value:

  - `image.scalar.add` `(I1 + r)`
  - `image.scalar.subtract` `(I1 - r)`
  - `image.scalar.inverse.subtract` `(r - I1)`
  - `image.scalar.multiply` `(I1 * r)`
  - `image.scalar.divide` `(I1 / r)`
  - `image.scalar.inverse.divide` `(r / I1)`
  - `image.scalar.equal` `(I1 = r)`
  - `image.scalar.not.equal` `(I1 <> r)`
  - `image.scalar.greater` `(I1 > r)`
  - `image.scalar.greater.equal` `(I1 >= r)`
  - `image.scalar.less` `(I1 < r)`
  - `image.scalar.less.equal` `(I1 <= r)`

  The following unary operations are also useful:

  - `image.negate` `(-I1)`
  - `image.not` `(NOT I1)`
  - `image.abs` `(absolute pixel values)`

- For neighbourhood operations between an image and a template:

  - `convolution`
  - `abs.convolution`
  - `multiplicative.maximum`
  - `multiplicative.minimum`
  - `additive.maximum`
  - `additive.minimum`
  - `median`

- For global operations on an image:

  - `sum.pixels`
  - `max.pixels`
  - `min.pixels`
  - `find.histogram`
  - `cum.histogram` (funds the cumulative histogram)
  - `scale.image` (scales an image to lie within given lower and upper bounds)
  - `all.pixels` (returns `TRUE` only if all pixels in the image are `TRUE`)
  - `any.pixels` (returns `TRUE` if any pixels in the image are `TRUE`)

6.5.1 Example image processing programs

To show how these routines can be used to put together a user program, consider the following two simple examples.

Thresholding an image Using a given threshold value as a cut-off point, pixels below the threshold are set to 0, while those above it are set to 255. To threshold an image `I1` to give an image `I2`, we use two of the above point-to-point procedures:

```
image.scalar.greater.equal (I1, Threshold, Temp)
image.scalar.multiply (Temp, 255.0 (REAL32), I2)
```

Edge detection A common algorithm for finding edges in an image is to use Sobel edge detection. This finds edges in a horizontal direction, then in a vertical direction, and computes a final combined edge strength which is then thresholded.

Finding the edges in one direction (vertically, say) can be done using convolution with the template shown in Figure 6.11. This looks for a difference between the grey levels on either side of a vertical line running through the middle of the neighbourhood. Horizontal edges are found using the template shown in Figure 6.12.

Using the standard method of representing templates outlined previously (row, column and weight for each defined position), we can declare and define two suitable template variables as follows:

```
[7][3]INT32 SV, SH:
...
SV := [[6,3,3],[-1,-1,-1],[-1,1,1],[0,-1,-2],
                [0,1,2],[1,-1,-1],[1,1,1]]
SH := [[6,3,3],[-1,-1,-1],[-1,0,-2],[-1,1,-1],
                [1,-1,1],[1,0,2],[ 1,1,1]]
```

The complete edge image can then be calculated using the following code:

```
abs.convolution (I1, SV, Temp1)
abs.convolution (I1, SH, Result)
image.image.add (Temp1, Result, Temp2)
image.scalar.greater.equal (Temp2, Threshold,
                                         Temp1)
image.scalar.multiply (Temp1, 255.0 (REAL32),
                                        Result)
```

6.5.2 Using libraries

To save having to recompile the entire suite of standard routines in both the communication and image processing layers every time the user's program is

–1		1
–2		2
–1		1

Figure 6.11 Edge detection using convolution to find vertical edges.

–1	–2	–1
1	2	1

Figure 6.12 Finding horizontal edges using convolution.

changed, it makes sense to put the standard procedures into a library. However, to be able to compile these routines separately, everything must be passed to the routines as parameters. This includes the eight channels which are accessed by the communication routines, and the node's row and column identity. Since the image processing routines themselves call the communication routines, these additional parameters must also be passed to them. This extra block of system parameters to every standard procedure disrupts the elegance of the coding somewhat, but is worth sacrificing in practice for speed of program development.

6.6 The future with the T9000

The programming guidelines given so far have been targeted very much to implementation on T800 transputers. With the T9000 now on the market, and occam 3 compilers and C104 message routing chips due in its wake, the outlook for transputer-based image processing is looking even more attractive. (It may take a little time, however, for T9000-compatible frame grabbers to become available.) While most of what has been said in this chapter will still be applicable to constructing well-structured image processing software on T9000-based systems, there are a few advances which can be exploited to good effect. Apart from the increased execution speed, the changes will mainly have the effect of simplifying software construction, and hiding some of the more tricky coding aspects.

With the T9000's Virtual Channel Routing capability (when combined with the C104 routing chip), the restriction to four links per processor is effectively done away with. It is therefore possible for each processor to have a link to all the other processors. In terms of our recommended approach to software development in this chapter, this architectural advance will greatly simplify the writing of the communications layer in our layered model. Explicit routing mechanisms for image distribution and border swapping disappear, and the coding becomes much simpler and more elegant. At the higher levels, however, it is still worth partitioning the image data, and holding segments (with replicated border regions) locally, in the same way as outlined in this chapter.

In occam 3, one new feature of the language which will have particular benefits for implementing our layered model is occam 3's library facilities. It will be much more convenient to put the various layers into libraries, with simpler interfaces to their users. This packaging facility is particularly necessary in parallel programming, where it is always desirable to hide communications in a separate layer. It is a pity it has taken so long to provide it.

6.7 Concluding remarks

We have outlined a strategy for the development of software for parallel low-level image processing which is not only straightforward and disciplined, but also has the important property that the software is *scalable*. More transputers can be added, and only one constant needs to be changed. It is vital to adopt this philosophy right from the outset. The approach advocated is also well engineered, by being layered. This separates different levels of activity, and presents a cleaner, more abstract view of the architecture to the applications programmer. Although it may seem that there is rather a lot of support software to develop before getting into some actual image processing, the recommended foundation is well worth laying. Disciplined effort at this early stage will pay for itself many times over in the future.

Bibliography

For more information about image processing (not necessarily parallel), the book by Low provides a good, readable introduction to a wide range of image processing operations. Gonzalez and Wintz is regarded as something of a standard text on image processing for the more dedicated.

For a more detailed treatment of specifically transputer-based parallel image processing, an excellent survey by Undrill can be found in the *Proceedings of the Transputer Symposium* (March 1992), edited by Jane, Fawcett and Mawby. A more wide-ranging coverage of various aspects of image processing on transputers is the book edited by Hugh Weber.

Gonzalez, R.C. and Wintz, P. (1987) *Digital Image Processing*, 2nd edn. Addison-Wesley, Reading MA.

Jane, M., Fawcett, R. and Mawby, T. (eds). (1992) *Transputer Applications – Progress and Prospects*. IOS Press, Southampton, UK.

Low, A. (1991) *Introductory Computer Vision and Image Processing*. McGraw-Hill, New York.

Weber, H. (ed). (1992) *Image Processing and Transputers*. IOS Press, Southampton, UK.

Chapter 7
Porting a Fortran Application to a Transputer Network

P. J. Morrow and D. Crookes

7.1 Introduction

This chapter provides a description of the porting and parallelization of an image processing application to a transputer network. The application concerned originally ran on a VAX 11/750 and Intellect 200 Image Processing system, and consisted of some 8000 lines of code. Much of the material presented here is based on a final report on this project (Crookes and Morrow, 1989) and a paper found in Morrow and Crookes (1991).

7.1.1 The application

The application to be parallelized is an image segmentation and analysis system for infra-red images of helicopters (Irwin and Wilkinson, 1988; Irwin, 1989). The scene depicted in such images is segmented and classified into a number of specific **objects** (for example, rotor, glass, target or cloud).

The interpretation process is implemented in two stages: initial segmentation (including feature extraction) and knowledge based classification. Image segmentation is achieved by assigning each pixel to one of a reduced set of allowable grey levels using a maximum likelihood operator. A figure of merit for the segmented image is calculated, based on maximizing the correspondence between the boundary pixels in the segmented image and the pixels of high gradient in the original image, whilst simultaneously minimizing the grey level difference between the original and segmented images.

A simple hill-climbing algorithm is then used to vary the positions of the allowable grey levels so as to maximize the figure of merit. When a local peak is reached, a number of features are extracted from the segmented image at various resolutions (for example, edge values, difference between the original and segmented images, connectivity).

A knowledge base containing *a priori* information about typical scene contents is then applied to the feature data. This is a *frame*-based process which uses high-level rules to ascertain quickly the broad nature of the segmented scene before a series of production rules are applied to provide a final scene classification.

7.1.2 The porting process

The original program was written in VAX Fortran, and took advantage of many non-standard VAX language extensions which were not available in the version used for the transputer implementation (3L Fortran). The style of coding also tended to be non-structured, perhaps because the program was originally experimental and not intended for further development. These factors mean that the program was probably typical of 'dusty-deck' Fortran programs in the real world, which have often lost whatever structure they initially had, and are usually thought to defy parallelization.

The porting process was carried out in three main stages, namely a single transputer implementation, a parallel implementation of the segmentation stages and an integration phase.

- Initially, a single transputer implementation of the whole system was developed. This involved (among others) modifying the program to resolve differences between VAX Fortran and 3L Fortran and also to recode the input and output sections (which used rather different hardware).
- A parallel version of the first (low-level segmentation) stages of the system was then developed to run on a pipeline of transputers. In addition to modifications to the system software itself, a number of runtime support routines also had to be written.
- Finally, the parallel segmentation stage and the remaining analysis stage were integrated together to produce a complete system running on the transputer network.

The process of porting and parallelizing this application to a transputer network is described in the following sections. Initially, a model of the existing software system is outlined. This is followed by a description of a parallel execution model which is based on observations of the initial model. The actual process of porting the software is then described, followed by an overview of the runtime support routines which were written. Finally, some results/conclusions are presented concerning the complete project.

7.2 Application software model

Two models of the system were produced. Initially, a model of the existing (sequential) system was developed, which was then analysed to allow a parallel software model to be produced.

7.2.1 System software model

The system consists of two main components, namely a low-level segmentation stage and a higher level analysis or classification stage. Initially, the segmentation algorithms are executed (for example, edge extraction, thresholding and histogram operations). Thereafter, the higher level analysis controls the execution of the system. It can request the segmentation process to be performed repeatedly thus **tuning** its output, which is then used in the classification process. Thus the software can be considered as having a pyramidal structure with data flowing to the analysis stage and control information being sent from the analysis stage to the segmentation stage. This is illustrated in Figure 7.1.

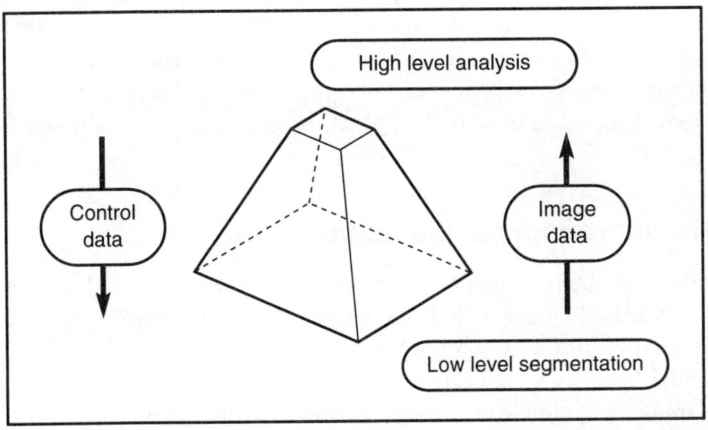

Figure 7.1 Pyramidal system structure.

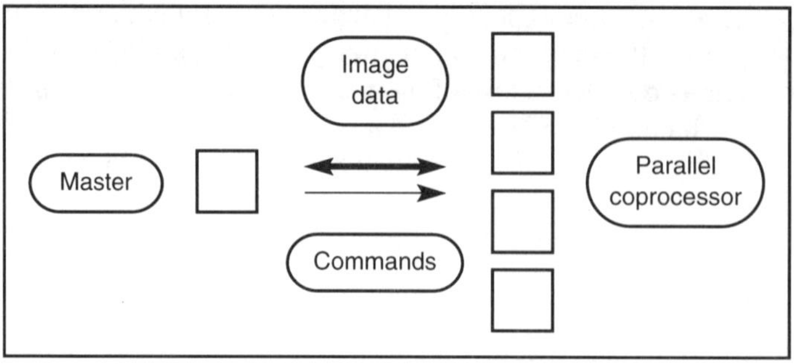

Figure 7.2 Parallel execution model.

7.2.2 Parallel execution model

On examination of the complete image segmentation and analysis system and its software model, it was obvious that the higher level analysis stage was predominantly sequential in nature. (Although the rule-based, higher level, interpretive processes could possibly be performed in parallel using a distributed database approach, it was not possible to investigate this within the time constraints of the project.) However, the segmentation stage could be parallelized and offered good potential for speed-up gains. Based on these observations, we developed a parallel execution model consisting of a master process (analysis) and a parallel coprocessor (for segmentation), as illustrated in Figure 7.2. (The coprocessor was configured as a simple pipeline.) The master process controlled the execution of the system and communicated all required image data to and from the coprocessor.

The segmentation stage involves a number of low-level image processing operations which can be considered as **array processing** operations. Thus it was obvious that by taking a data parallelism approach to the problem significant speed-ups could be achieved. Each image was split into sections and distributed over the available transputers in the coprocessor network. Each transputer runs the same program, only processing its own local portion of the image. Any scalar data used in the system is replicated over all processors.

7.3 Single transputer implementation

As a preliminary step in the porting process the system was initially ported from the VAX to a single transputer. The main reason for this was to allow us to identify incompatibilities between the version of Fortran which had originally been used for the system (VAX Fortran) and 3L Fortran which was to be used in the parallel implementation. Many of the problems at this stage were caused mainly by a lack of structure in the original Fortran code and poor programming practices, and involved substantial changes to the original source, which required a significant amount of time and effort. However, the majority of them arose from the use of non-standard FORTRAN 77 facilities provided on the VAX computer.

In addition to changes due to incompatible versions of Fortran, some further modifications were required in relation to Input/Output. The image display routines used in the original program were obviously tailored to the Intellect Image Processing system and had to be re-coded for our display board.

Finally, some memory limitations were encountered in the single transputer implementation due to the large number of 2- (and 3-) dimensional arrays being used. To reduce the static storage requirements, a number of these arrays were placed in common blocks.

7.3.1 Incompatible versions of Fortran

The changes which were required when moving from VAX Fortran to 3L Fortran are summarized below.

Integer type sizes In VAX Fortran integers can be declared as being of size 2 or 4 (or 8) bytes, whereas the version of 3L Fortran which was used was limited to integer sizes of 4 bytes. All of the arrays used in the original program to store images were of type Integer*2. As these arrays were relatively large in size, their conversion to arrays of integers of four bytes in size led to a substantial increase in static storage.

Structured data types VAX Fortran allows structured data types. For example, the following structure taken from the program defines a 'blackblob':

```
structure / blackblobtype/
  integer currentlev,numcon,iblobcord(64),jblobcord(64)
  integer goodconnect,poorconnect
end structure
record/blackblobtype/blackblob
```

This construct is not available in standard FORTRAN 77, and in the conversion process each of the fields of these record structures had to be redeclared as an individual variable (each field name is appended to the structure name). This resulted in a considerable amount of editing, and care was needed in the indexing of fields which were declared as arrays. In the example code above, `blackblob` was later defined in a common block called `blackblob` as a 2-dimensional array `blackblob(3,64)`. The single dimensional array field `iblobcord` therefore had to be declared as a 3-dimensional array in the transputer implementation. Therefore every instance of:

```
blackblob(x,y).iblobcord(z)
```

was replaced with:

```
blackblobiblobcord(z,x,y)
```

where blackblobiblobcord had previously been declared as a variable.

This process may seem to produce very long identifier names; however, it was felt that unless a well-structured method of editing was introduced, errors could easily occur.

Non-integer array subscripts Non-integer array subscripts are automatically converted to integers in VAX Fortran. In the original source, many of the variables used to subscript arrays were implicitly defined as reals. To convert to standard FORTRAN 77 such subscripts were converted to integers using the **int** function which truncates any fractional parts.

Control transfer into conditional blocks VAX Fortran allows goto statements directing control into IF, ELSEIF and ELSE blocks. Unfortunately, control transfers of this kind appeared frequently in the original source. To comply with FORTRAN 77 requirements, and at the same time retain the program logic, the blocks of code to which the goto statement passes control were repeated at the point where the goto statement occurs, and the goto statement removed. In some cases, this involved repeating a considerable amount of the code. However, it was felt that this was the easiest method to choose, since a considerable understanding of the system was required for any other method.

Control transfer into iteration blocks In the original source the following section of code typically appeared at several points:

```
      do 10 ...
         do 20 ...
            {statements 1}
20       continue
10    continue
      {statements 2}
      goto 20
```

The 3L Parallel Fortran compiler signals an error at any attempt to pass control into the range of a do loop. This meant that the original code had to reorganized in many cases.

Default file units VAX Fortran allows file units to be declared by default within a READ or WRITE statement by using the following format:

```
write (1, *) 'blob = '
```

By default, if unit 1 is not open when such a statement is executed, the data will be assumed to be in a file named FOR001.DAT. With 3L Fortran, an OPEN statement of the following format is required to declare such a file:

```
open (unit=l,status='scratch',access='sequential')
```

Initialization of array variables When using VAX Fortran, array variables are initialized to zero or false. VAX Fortran programmers often rely on this. However, this is not the case with 3L Fortran, and many of the arrays in the original program had to be explicitly initialized.

A number of other problems occurred during the porting of the program, but those outlined above cover the main class of problems encountered. As should be clear from the above, the standard of programming practice used to construct the original program left a lot to be desired (however, the program was developed in an experimental context, and was not intended for future development). This programming *style* presented many difficulties, and dealing with them proved to be one of the more time consuming aspects of the project.

7.3.2 Image input/output

The image display routines used to display the various images produced by the system had to be rewritten in occam to produce image displays on a B007 transputer graphics board. As all of the image display routines used in the original program were written in Fortran, care had to be taken to ensure that the occam tasks were dealing with the image as originally intended. One notable difference between the customary ways in which these different languages deal with images is the way in which they reference the coordinates of the images. Whereas occam uses a row/col matrix, Fortran uses an x/y matrix. Compensating for this difference proved important not only for displaying the image, but also in reading and processing the image. An important part of the image recognition section of the program involved assessing the size (width and height) of parts of the image. Dealing with the coordinates of the image incorrectly here would have led to misleading results.

7.3.3 Memory limitations

Memory space requirements proved to be an important consideration in the porting of the program. A B004 transputer board with 2 Mbytes of external memory was being used. The compiled/linked system itself took up a considerable portion of the available memory. On initial attempts to run the system it either hung or an error message indicating lack of static storage space was given. This space problem resulted from the considerable number of 2- and

3-dimensional arrays used in the program. In an attempt to reduce the amount of static space being used, array variables which could have been placed in common blocks but had not, were subsequently placed in common blocks. This simple refinement succeeded in overcoming many of the memory problems associated with the single transputer implementation.

7.4 Parallel segmentation

On examination of the complete image segmentation and analysis system it was decided that the most appropriate parts to parallelize were the segmentation stages. (Initially, this was all that was to be parallelized, although ultimately it was integrated with the analysis stage to provide a complete parallel version of the system – see Section 7.5.) The segmentation stages involve a number of low-level image processing operations which can be considered as array processing operations. The segmented data produced by this stage is used by the analysis section to produce a final classified image. It was obvious that to get the best performance from a parallel architecture for segmentation we should use a data parallelism approach to parallelization. The strategy we used for parallelizing the system was to split up each image used and distribute them over the available processors – one strip or section per processor. Modifications are also made to any image declarations. This segmentation stage is then run on each of the transputers in a pipeline. Each transputer runs the same program, only processing its own local portion of the image. Scalar data which is independent of image values is replicated over the pipeline where a copy of the data is held on each transputer. Special cases where scalar variables are calculated from distributed image data must be considered separately. In these cases, scalar calculations are carried out on local image sections and then global values are calculated using these local results.

The process of parallelizing the system involves identifying operations which depend upon image data. If an operation involves processing a distributed structure (that is, an image), then modifications are made to the Fortran Do-loop bounds to match the image section size. To handle image Input/Output between the frame-store and the pipeline, a set of runtime support routines were constructed.

The problems associated with moving from one transputer to n transputers are dealt with in detail in the following sections, and cover the areas of image declarations, image Input/Output, processing of images (array bounds) and global processing of distributed data. (An overview of the set of runtime support routines is presented later in the chapter.)

7.4.1 Image array declarations

Each image used in the program must be divided into a number of strips, with each processor being allocated a strip to process. The size of the image section therefore depends both upon the image size and the number of transputers in the pipeline. An image declaration in the sequential program would be as follows:

```
Integer EdgeImage (128, 128)
```

whereas in the parallel implementation this declaration must be modified to:

```
Integer EdgeImage (SectionRows, SectionCols)
```

where SectionRows and SectionCols are predefined constants held in a global common block, and represent the number of rows and columns in an image section. For example, for four transputers and an image of 128 × 128 pixels, the size of each segment would be (34, 128). The value of 34 (rather than 32) allows for a border around each segment.

7.4.2 Image input/output

The images to be processed by the system must be read from disk or captured from an image frame grabber. The sequential version of this may be:

```
Call ReadImage (Image, 'File1.Img')
```

The subroutine ReadImage reads data from the file and stores the whole image in the variable Image. However, for the parallel implementation, each node in the pipeline must execute the following two subroutine calls:

```
Call NodeReadImage (PartImage, 'File1.Img')
Call SwapSectionBorders (PartImage)
```

The routine NodeReadImage communicates with the host machine via the master processor and will read the image from the file and automatically distribute appropriate image sections to each of the transputers. The second routine is invoked to exchange section border information between adjacent transputers, and is done for efficiency since many image processing operations require access to a pixel's neighbourhood. (This operation could be performed as part of NodeReadImage; however, it is necessary to have it as a separate routine since border information nust be exchanged at other times, such as each time an array section is processed.)

7.4.3 Simple processing

The modifications required when an image array is being processed are quite straightforward. They simply involve editing the original bounds of Do-loops to match the size of local image sections. For example, the following piece of code from the original program modifies an image called Image:

```
Do j = 1, 128
   Do i = 1, 128
      Image (i, j) = ...
   End Do
End Do
```

The equivalent code for the parallel implementation is obtained by simply modifying the loop bounds as follows:

```
Do j = StartCol, EndCol
   Do i = StartRow, EndRow
      Image (i, j) = ...
   End Do
End Do
```

where the variables StartRow, StartCol, etc. depend upon the image and pipeline sizes, and are predefined in a global common block.

7.4.4 Global processing of distributed data

In many instances in the sequential program scalar (non-image) variables are evaluated using operations which depend upon the whole image (for example, finding the histogram of a distributed image). Obviously, the correct value for these variables can only be obtained from examining the whole image. In the pipeline implementation, each node will calculate a value which depends only upon the image section which it holds. Subsequently, some sort of global evaluation must be performed. This is once again achieved by calls to the runtime support software. For example, consider the following piece of code, taken from the sequential program, which finds the histogram of an image:

```
Do j = 1, 128
   Do i = l, 128
      Histogram (Image (i, j)) = Histogram (Image (i, j)) + 1
   End Do
End Do
```

The equivalent code for the parallel implementation is:

```
Do j = StartCol, EndCol
   Do i = StartRow, EndRow
      Histogram (PartImage (i, j)) =
         Histogram (PartImage (i, j)) + 1
```

```
        End Do
    End Do
    Call GatherScatterHistogram (Histogram)
```

The variables StartRow, StartCol, etc. again depend upon the image and pipeline sizes, and are predefined in a global common block. The only other significant change to the original code is the addition of a subroutine call subsequent to finding the local histogram. Each node invokes this routine, which takes care of gathering together all the local histograms, calculating a global one and scattering it back to each node. Other examples of local scalar operations are finding minimum and maximum values and summation over an image.

7.5 System integration

As indicated earlier, it was initially intended only to parallelize the segmentation stage of the system. However, once this was achieved we decided to integrate the parallel segmentation program with the analysis section to provide a complete system running on a transputer network. In considering this it was recognized that the segmentation section effectively acted as a *slave* to the analysis section. Segmentation of an image is performed giving an initial segmented image with only 2/3 grey levels. Analysis is then carried out on this image, and if this process does not produce satisfactory classifications (according to some predefined criteria) the segmentation levels may have to be retuned. Hence the segmentation stage is recalled to provide new segmentation levels. This retuning process continues until a satisfactory analysis is obtained.

In the parallel execution model for the system outlined previously, the analysis stage effectively became the whole sequential program, with the parallelized segmentation removed. In its place a number of commands are issued to the (slave) coprocessor, which will then execute the segmentation process in parallel. In addition to commands being sent to the segmentation stage, any required image data must also be transferred with any results being returned. This process is illustrated in Figure 7.3.

The analysis section controls the overall operation of the system. The segmentation process is called to perform an initial segmentation of the input image. Due to the sharing of image (and other) variables between the two stages via the use of common blocks, this information must be transferred to the parallel coprocessor prior to the segmentation process being executed. While segmentation is being performed, the analysis process acts as a server to it to allow text and image I/O to and from the processor pipeline. Once segmentation is complete the common data is transferred back to the analysis stage and the segmented image is analysed. If this analysis results in an acceptable classification of the scene depicted in the image (according to a set of rules defined in a rule database), then the system terminates. Otherwise, segmentation is repeated again and various parameters and segmentation levels are retuned to new values until an acceptable scene classification is achieved.

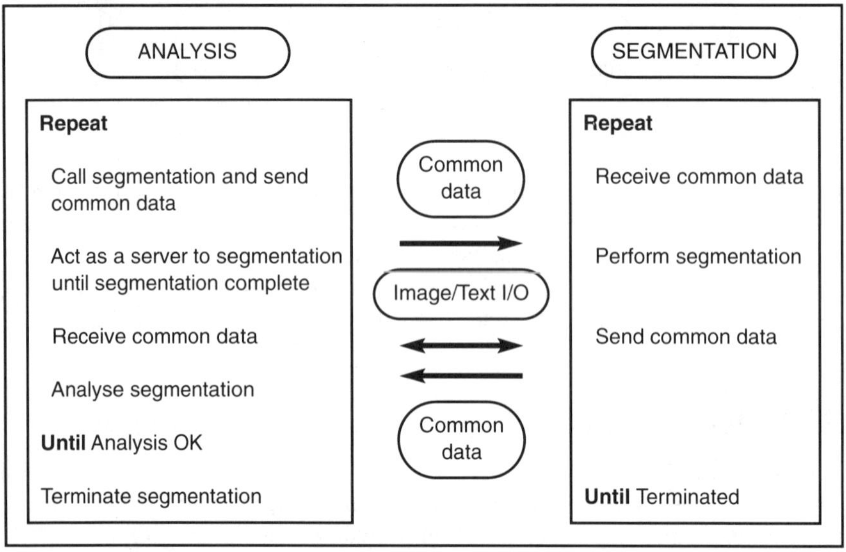

Figure 7.3 Analysis and segmentation integration.

There were two main problems associated with the above method of implementing the complete system on the transputer network:

(1) *Common data and shared variables* A number of common data sections were used in both the segmentation and analysis parts of the system and, in addition, some variables (not in common data sections) were also shared between them. Subsequently, if modifications were made to these shared variables in either the segmentation or analysis stages, the new updated variables must be communicated to the currently idle stage. This was implemented by a set of complementary runtime support routines to send/receive common data (see control flow in Figure 7.3.

(2) *Image and text input/output* When the segmentation stage is running, image data or text messages may need to be displayed on either the frame-store or PC monitors. Once again, to allow this two sets of runtime support routines were written. One set runs on the pipeline and effectively makes requests to the host (such as *display the following image*, or *print the following message).* The other complimentary set runs on the host and deals with the requests coming from the pipeline.

Once the runtime support routines had been written and incorporated into the system, little further effort was required to obtain a working system.

7.6 Runtime support software

In an attempt to keep changes to the original program to a minimum, a large number of runtime support routines were written. (These routines have been mentioned in previous sections.) They can be classified into the following categories:

- image input/output – allows the pipeline to receive/transmit image data from/to a controller process;
- text input/output – allows nodes in the pipeline to communicate messages with the host computer;
- scalar input/output – allows communication of result/diagnostic values to and from the host computer;
- gather/scatter operations – for example, histograms, minimum/maximum values or summations;
- miscellaneous – for example, Sobel edge detection algorithm (designed for a pipeline).

Once this library had been developed the parallelizing operation was relatively easy. It consisted of simply identifying those areas of the program which either performed operations on images or calculated scalar values which depended on image data. The appropriate code was then replaced with calls to the runtime library routines.

7.7 Results and timings

Although a detailed study of the performance of the ported software was not the main aim of this study, some timings of the system's operation were carried out. Six test images were used (A–F), since the nature of the image influenced the number of times the low-level segmentation was called. First, the complete single-transputer version was run (on a T4 transputer). The first column of figures in Table 7.1 gives these timings (in seconds). The system took approximately five minutes to execute. The next phase was to take the parallel segmentation stage, and to time it on different pipeline lengths (again using T4 transputers). This stage is actually called several times (anything up to eight or nine times) during the course of the complete program's execution; but the timings in Table 7.1 are for just a single call of the segmentation process (again, on six different images). As can be seen, significant improvement was obtained up to around eight transputers, after which communication costs reduced the efficiency.

Table 7.1. Parallel segmentation timings for different pipeline sizes.

Image	*(Full version) Sequential(s)*	*(One iteration of low level segmentation)(s)* 2	4	8	16
A	(260)	20	12	9	7
B	(255)	40	23	15	10
C	(310)	25	15	12	9
D	(330)	27	15	8	6
E	(285)	20	12	8	6
F	(210)	31	17	10	8

For the complete system, significant improvement was obtained in going to a four transputer pipeline, but there was little gain in using more transputers. The main reason for this is concerned with the amount of communication of data which is required between the segmentation and analysis stages. This data consists of a number of images and various other shared variables. Thus, four transputers seems to be a reasonable upper limit for this *coprocessor* approach for this particular application. The routines which implement this transmission of data could be optimized further with an associated improvement in the results.

We further found that, running the complete single-transputer version on a single T8, the performance improved by a factor of approximately 2.5 over the equivalent T4 implementation. However, we would expect that on a pipeline of four T8 transputers, performance would not necessarily double as it did for T4s, since the communication bottleneck would occur at an earlier stage.

7.8 Conclusions

Some conclusions are now drawn concerning the complete project and the problems associated with porting a large image processing program on to a transputer network.

7.8.1 Development time

The most time-consuming parts of the porting process were, in decreasing order of time required:

(1) Making the necessary modifications to deal with the different versions of Fortran. However, it is felt that the particular system considered in this project would not be typical.

(2) Writing the runtime support for the parallelized version (in particular for the segmentation stage).

(3) Modifying the Input/Output sections of code to handle a change in the display hardware used.

(4) Modifying the low-level segmentation code for parallel execution.

(5) Identifying and implementing the master/slave split when producing the full parallel system.

If another porting process were now to be carried out on a similar image processing application, it is felt that the process would be much quicker. The following points, which mirror those above, are relevant in this regard:

(1) If the code was written in standard Fortran (even if poorly written), then the most time-consuming phase could practically be eliminated.

(2) The existing runtime support which we have developed would be reusable (for the lower level image processing operations). It has been coded as a general communications package, and the details of what goes on behind the scenes can be hidden from the user.

(3) Any image processing application is inherently susceptible to a change in Input/Output hardware. However, we have developed a set of Input/Output routines which could again be reused.

(4) Phases (4) and (5) above were relatively quick, though they were definitely intellectually more taxing. They also required a certain degree of understanding of the original program: for instance, to identify which images in the program should be distributed, scalar data to be replicated, etc.

7.8.2 Speed-up

At the low-level segmentation stage, almost linear speed-up was achieved (at least up to eight transputers). However, the integrated multi-level system (master/slave) did not achieve such good results. There are two main reasons for this. First, and obviously, the analysis stage was not parallelized, which means that the pipeline is idle for periods during this stage. Second, the communication of image data between the segmentation and analysis stages contributed greatly in reducing the efficiency of the overall system.

Time did not permit an attempt to parallelize the medium- and high-level operations in the analysis stage. Since much of the medium-level analysis involves reducing the data to, say, 8×8 arrays, this is amenable to parallelization by distributing these. Much of this analysis could then be done on the coprocessor pipeline. This would reduce the bottleneck of communication between the controlling processor and the pipeline, so the current effective limit of four transputers could be increased. Also, at the higher level analysis stage, the data has been reduced in volume, so should not be so susceptible to the communications bottleneck.

7.8.3 Understanding of the original program

A certain amount of understanding of the original program was necessary to identify the low-level segmentation (which could be put on the slave pipeline), and to identify what data had to be transferred between master and slave, etc. Fortunately, the original programmer was available for consultation during these phases of the project. Without this, the task would have been significantly more difficult.

Finally, one particularly successful aspect of the project is that it has demonstrated the feasibility of taking a large, poorly-structured Fortran program, and implementing it on a transputer network. It may be more straightforward to parallelize a carefully-worked and well-structured program; but 'dusty-deck' Fortran programs in the real world have often lost whatever structure they initially had, and are usually thought to defy parallelization. For one such dusty-deck program, this project has achieved a reasonable speed-up using a small number of transputers, and we have reason to believe that the coprocessor technique could produce further speed-up with a larger number of transputers. The project's achievement in this regard is significant.

Acknowledgement

This work was supported by an SERC Transputer Initiative research grant. The work was carried out through the Northern Ireland Regional Transputer Support Centre, and the use of the Centre's facilities is gratefully acknowledged. A considerable amount of help and support was given by Short Brothers plc, who supplied the original program.

References

Crookes, D. and Morrow, P.J. (1989) *A study in multi-level image processing on transputers.* EMR Final Report, January.

Irwin, P.D.S. (1989) *Real-time processing techniques for infra-red image understanding.* PhD Thesis, Queen's University of Belfast, Northern Ireland.

Irwin, P.D.S. and Wilkinson, A.J. (1988) An integrated image segmentation/image analysis system. In: *BPRA 4th International Conference on Pattern Recognition*, Cambridge, UK, March.

Morrow, P.J. and Crookes, D. (1991) *Parallelising an image segmentation and analysis system for infra-red images.* In: *Applications of Transputers 3*, (Durrani, T. *et al.*, eds.), Vol. 2, pp. 327–332. Amsterdam: IOS Press.

Chapter 8

Physiological Modelling of the Human Auditory System

K. Adamson

8.1 Introduction

This chapter presents an approach for developing the concurrent structure of a program in the domain of physiological modelling, namely the modelling of the human auditory system. One model which exhibits a degree of natural concurrency is considered. The usual form of electrical representation, the digital filter diagram, is taken as the specification from which a concurrent model is developed. This specification (final digital filter representation) is introduced, and the derivation of the concurrent program is demonstrated.

Development in VLSI technology and the proliferation of parallel machine components suggest that parallel processing will have a major future impact in the design of computing systems. This is likely to be true in the field of physiological modelling, which tends to involve extensive use of signal processing techniques. The systems inherently have the requirements of very high performance with the added factor of operation diversification.

Two classes of solution prevail; either design specialist hardware, or employ current multiple processors. While specialist hardware may be an attractive solution for some applications, there is difficulty in those situations where the flexibility of a computer-based solution is preferable. With the multiple processor approach a major aspect of the design is the interconnections between processors. Typically, contemporary machine designs have used a bus for this, sometimes even the bus that the processor is using to fetch executable instructions and associated data. Regardless, a bus-oriented multiprocessor has limitations that render it quite inappropriate for the scale of systems that will need to be implemented. Single-bus multiprocessors demonstrate this by providing successively less incremental computational capability with the addition of successive machines.

In an attempt to address these problems, vendors have created a number of special purpose bus standards; one for memory, I/O, interprocessor communication, and so on. These measures simply build on the underlying weakness of a single-bus approach (even if dedicated to separate functions), and are ill disposed to real-world systems where locality is likely to exist.

The Inmos transputer system architecture (Inmos Ltd, 1988a) is an efficient implementation of addressless point-to-point interprocessor links. This promises to significantly simplify the design of multiprocessor machines, as an electrical connection between corresponding pins allows the creation of arbitrarily large systems. The use of occam (Inmos Ltd, 1988b) also assists the modelling process, as a concurrent occam program, to be targeted eventually to a multiprocessor configuration, will execute correctly on one processor, allowing simulations to be performed. There is, however, an urgent need for practical methodologies to facilitate the design of concurrent algorithms, and to enable the evaluation and use of parallel processing.

In general, the production of an effective concurrent design will require a reappraisal of the original problem. An initial temptation with transputer-based systems may be simply to convert to occam an existing design coded in a high level language, and to use the transputer as a new fast processor. It is unlikely that basing a concurrent design upon a sequential design will result in a clear or efficient solution able to capitalize on the advantage of parallelism. Adopting such an approach would complicate the design process as a consequence of the number of abstractions based on sequential programming introduced between the original problem and the concurrent solution.

8.2 Physiological models

The main emphasis of the work presented here is in the application of concurrency to the area of biomedical analysis. One physiological model which accurately matches the lower auditory functions has been developed to date (Adamson and Black, 1986). The physical model developed of the human basilar membrane is intended to simulate the natural biological system as closely as possible from the acoustical, physiological and clinical points of view. With this model our attention is directed towards the possible applications in assisting individuals with auditory impairments, and in providing essential understanding and information for those concerned with the treatment of such impairments.

The model simulating the lower auditory functions may be able to make useful contributions in the treatment of the hearing-impaired people. An area in which there is scope for considerable improvement is in the choice of a particular hearing aid for a particular patient. At the present time there is a huge variety of hearing aids available, each of which has its own particular characteristics such as high or low gain, wide or narrow frequency response, a frequency response tailored to the subject's audiogram, directional or nondirectional microphones, and so on. Unfortunately, there is a paucity of systematic research on which type of aid is best for which type of hearing disorder, and the 'prescription' of a hearing aid is often made by inspired guesswork rather than on the basis of sound scientific evidence.

Hearing losses may be broadly categorized into two main types. The first type, **conductive deafness**, occurs when there is some defect, usually in the middle ear, which reduces the transmission of sound to the inner ear. For example, a growth may occur on the eardrum, or the bones in the middle ear may ossify or stick together. In general, this results in a more or less uniform hearing loss as a function of frequency, and the difficulty experienced by the sufferer can be well predicted from the threshold elevation. A simple hearing aid is usually quite effective in such cases, and surgery can also be effective. The second type of hearing loss is called **nerve deafness**, or **sensorineural hearing loss**. Usually, the extent of the loss increases with frequency, but the difficulty experienced by the sufferer is not always well predicted from the audiogram. Sensorineural hearing loss may arise as a result of defects in many different parts of the auditory system, and the particular difficulties experienced by the sufferer, and the types of symptoms exhibited, depend upon which part of the system is affected. Patients often have difficulty in understanding speech in noisy environments, and the condition is not completely alleviated by a hearing aid. Most sensorineural losses cannot be treated by surgery.

An example of an area in which our proposed transputer-based system may be useful is found in attempts to compensate for the loudness recruitment which commonly occurs in hearing losses of cochlea origin. In a recruiting ear the absolute threshold is elevated, but the level at which sounds become uncomfortably loud may be almost normal; the growth of loudness with increasing intensity is more rapid than normal. A hearing aid which amplifies all sounds equally is not satisfactory in such cases. If the gain of the aid is set so as to make faint sounds audible, then the more intense sounds will be overamplified, and will be painfully loud. One method of compensating for recruitment is to use a hearing aid which compresses the dynamic range of the input, so that low-level sounds are amplified more than high-level sounds. One drawback in this solution is that the amount of recruitment is usually not constant as a function of frequency. Thus the compensation appropriate at one frequency will be inappropriate at other frequencies. The underlying auditory model in our system uses a bank of 128 digital filters which splits the input into a number of frequency bands; it would therefore be possible to apply the appropriate amount of compression to the output of each filter. It is likely that in a wearable aid it would not be necessary to use as many channels. Adequate compensation for recruitment may be achieved with fewer channels, and there may indeed be a disadvantage in using a large number of channels, since the compression will tend to reduce spectral contrasts in complex stimuli.

A second way in which the preprocessing of sounds could be useful is in reducing the amount of background noise or reverberation. A common experience of hearing-impaired people is that they can understand speech well in a quiet situation with one speaker, but that understanding is difficult when there is background noise or reverberations. There are several computer based schemes for reducing background noise and/or reverberation, but they all have drawbacks and limitations, and none has been implemented, as yet, in a wearable aid.

Cochlear implantation is a relatively new and exciting area of research aiming to bring help to those people who are totally deaf through disorders of the cochlea. The number of people in this category is not clearly known, but may be considerable, and they cannot hear anything using a conventional hearing aid. In a large proportion of these the disorder is in the sensory organ rather than in the central nervous system, and the auditory nerve is partially intact. In such people it may be possible to create a sensation of sound by direct electrical stimulation of the auditory nerve.

8.3 Structure

In any imperative programming language, the structure of a program plays an important role. This applies to concurrent programming languages no less than it does for sequential programming languages, though unlike the case in sequential programming, program structure is less well understood. The lack of general agreement concerning how a concurrent program should ideally be structured acts as an impediment to the application of parallelism in many areas where benefits may be realizable.

Although considerable research effort is being directed at developing systematic approaches to concurrent program design, it will be some time before this permeates general application domains. Physiological modelling is one such domain where an evaluation of parallelism would be useful. This work is not meant to be a substitute for other more rigorous work, but merely to allow those with little experience or knowledge of concurrent programming to proceed towards evaluating parallelism in their own discipline. The rules are intuitive and only apply to systems based on digital filter diagrams, which excludes many areas but nevertheless embraces a fairly significant body of applications.

The rules are intended to simplify the construction of the concurrent structure of a program by identifying key features of digital filter diagrams, and establishing corresponding software analogues so that a straightforward mapping from one to the other can be made. Adopting this approach is unlikely to lead to an efficient or indeed pleasing structure; much more important within our objectives is the extent to which this can be rendered mechanical.

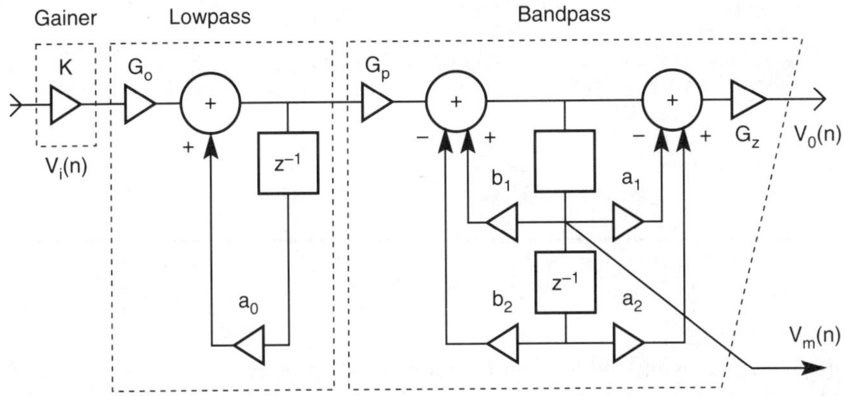

Figure 8.1 Basilar Membrane model performance.

From Figure 8.1, the key features are quite obvious, highlighted by the different kinds of symbol used. These are the GAIN, ADDER, DELAY and JUNCTION, indicated by a triangle, circle, square and interconnecting lines, respectively, embellished as appropriate. Thus in Figure 8.1, reading from the left, there are two GAINs in series, followed by an ADDER whose output is directed along two separate paths by a JUNCTION, one of which is into a DELAY succeeded by a GAIN, which is then fed into the ADDER already mentioned.

Each symbol has a specific function. A GAIN simply multiplies a value passing through it by some designated value, such as by K for the first GAIN in Figure 8.1. The multiplying factor may be positive such as for K, or negative such as for b2 in the same figure. An ADDER, as its name suggests, adds together the values present on its input lines, typically those entering from directions other than the right which usually indicates the output line, and passes the sum along its output line. A DELAY serves to suspend the progress of a value until the next or subsequent values are available. Finally, the JUNCTION merely disperses a value along two or more lines, greatly extending the complexity of systems that may be described.

Although the semantics of each symbol is simple, complex systems can be described. Furthermore, the meaning of each symbol is independent of their particular use, allowing them to be freely utilized in any arrangement. Both of these attributes, that is, simplicity and independence, are particularly important in that they allow the rules to be clearly related to specific symbols and similarly their software analogues.

8.4 Rules

The first general rule is that all of the features identified in the filter diagram become distinct procedures in occam. There will therefore be four types of procedure: GAIN, ADDER, DELAY and JUNCTION. These can be thought of as

```
PROC Gain (CHAN OF REAL32 input, output)
  VAL REAL32 coefficient IS 1.0:
  REAL32 value:
  WHILE TRUE
    SEQ
      input ? value
      output ! value * coefficient:
```

Figure 8.2

software components which can be substituted for their filter diagram equivalent. All of the software components must, of course, allow for flexible interconnection, so besides the implementation of their functionality, provision must exist for input and output lines.

In occam the interconnections will be implemented by channels; however, each software component need only address its own specific requirement. The interconnection stage is a separate and distinct stage related to a system, about which the software components are ignorant.

8.4.1 GAIN component

The occam implementation of the GAIN is shown in Figure 8.2. Two connections are permitted with other components, identified in Figure 8.2 as ‘input’ and ‘output’. Their type is designated as `REAL32`, since the models developed from filter diagrams operate on floating point numbers, and in the two models investigated, 32-bit real numbers have proved adequate. The actual value (coefficient) to be applied on values passing through the GAIN is declared within the procedure, in addition to a variable into which the incoming value is read. The body of the procedure is an infinite loop inside which an input value is read, multiplied by the coefficient and output.

8.4.2 ADDER component

An ADDER is similar to the GAIN in so far as it inputs values, modifies them and outputs a single value. However, it is significantly different in that there must be more than one input line. This is reflected in Figure 8.3, where two input lines are implemented.

Like the `Gain` procedure, the `Adder` body is an infinite loop. In fact, each iteration of the body only completes if input data is available, and whenever this is not the case the ADDER must wait. Conveniently, the input and output processes in occam implement this, suspending a process awaiting input until it duly occurs.

```
PROC Adder (CHAN OF REAL32 input1, input2, output)
  REAL32 value.1, value.2:
  WHILE TRUE
  SEQ
    PAR
      input1 ? value.1
      input2 ? value.2
      output ! value.1 + value.2:
```

Figure 8.3

A significant feature of the `Adder` procedure is its use of the `PAR` constructor to enable the successful collection of both inputs before proceeding to calculate the sum and output it. Accepting both inputs in parallel guards against introducing deadlock into the component, allowing the components connected to the 'other' end of the input channels to present their data in any order. If these inputs were done sequentially, then the component would have to force a specific order upon the connected components *vis-à-vis* the outputting of data, thereby compromising the independence of components.

8.4.3 DELAY component

Implementing a DELAY in an environment where there is no global synchronization appears at first glance somewhat difficult. However, when account is taken of the synchronous communication properties of occam, the solution is quite straightforward. This can be illustrated by considering at what point the DELAY must be effective, as distinct from how it is implemented. In Figure 8.1 the ADDER marks this point, where the expectation is that one input line carries the *N*th value and the other the delayed and modified *N*–1th value. The Lowpass section of Figure 8.1 clearly demonstrates this, and from the implementation of the ADDER component in Figure 8.3, all that need be ensured is that when the ADDER inputs the nondelayed value, a delayed value will also be available, eventually. It can be seen that the `Adder` procedure effects the synchronization, only allowing execution to proceed towards the actual adding when both delayed and nondelayed values are present. The order in which these values arrive at the ADDER is unimportant as a consequence of the `PAR` constructor. This allows the delayed value to present itself first, without in any way compromising the digital filter design.

Implementation of the `Delay` procedure need not, as outlined above, address the problem of synchronization of delayed and nondelayed values as this is already incorporated in the `Adder` procedure. This is important in that it preserves the independence of the DELAY component and makes its implementation simple, as shown in Figure 8.4.

```
PROC Delay (CHAN OF REAL32 input, output)
  REAL32 value:
  SEQ
    value := 0.0
    WHILE TRUE
      SEQ
        output ! value
        input ? value:
```

Figure 8.4

The `Delay` procedure merely inputs a value and then outputs it, having in the initial case output a zero value. An initial zero output is the means by which the DELAY is introduced, guaranteeing that when an `Adder` procedure initially inputs values on its two incoming channels, execution will occur without the delayed value having any effect on the computed output. This is consistent with the semantics conveyed by the digital filter diagram. On subsequent inputs an `Adder` procedure will receive delayed values related to the actual processing of whatever model is in place. For example, referring once again to the Lowpass section of Figure 8.1, it is evident that when the `Adder` procedure inputs the second value on the nondelayed channel, the previous value through the ADDER (first value) will present itself, albeit modified by a `Gain` procedure, on the delayed input channel.

Since no modification to the value is made, a temptation might be to remove the `Delay` procedure and embed the initial zero in the `Adder` or `Gain` procedure. Whilst this ultimately may be advisable in achieving performance improvement, this course is rejected at this stage of development on the grounds of simplicity and clarity.

8.4.4 JUNCTION component

The JUNCTION component poses no problems in its implementation and can be cleanly mapped to occam. As shown in Figure 8.5, a value is input and dispersed on whatever number of output channels are necessary. Tailoring the procedure to disperse on *N* channels simply requires the declaration of the requisite channels, and the insertion of an output process for each of these channels within the `PAR` constructor. Figure 8.5 illustrates the procedure suitable for the JUNCTION component in the Lowpass section in Figure 8.1.

Like the `Adder` procedure, `Junction` employs the `PAR` constructor to concurrently output values on separate channels. Besides maintaining an analogue of the semantics of a JUNCTION in a filter diagram, the use of the `PAR` constructor prevents deadlock from occurring. Deadlock is obviously more difficult to prevent where a procedure has many interconnections, and since this could be particularly troublesome with the JUNCTION component (due to the variability of output lines), the use of the `PAR` constructor is essential.

```
PROC Junction (CHAN OF REAL32 input, output.1,
                                     output.2)
  REAL32 value:
  WHILE TRUE
    SEQ
      input ? value
      PAR
        output.1 ! value
        output.2 ! value:
```

Figure 8.5

8.5 Systems from components

Throughout the derivation of the `Gain`, `Adder`, `Delay` and `Junction` procedures, care has been taken to avoid establishing relationships between them which would impede their use in the development of a system. The construction of the concurrent structure of a filter diagram can thus be achieved by selecting the appropriate procedure corresponding to a component, and connecting it as necessary.

The connection of procedures is achieved by declaring channels and matching them appropriately so as to model the lines on the filter diagram. Accordingly, there will be a distinct channel required for each line between any two symbols in the filter diagram. Taking the Lowpass section of Figure 8.1, seven channels are necessary. These are between GAIN(K) and GAIN(G_o), GAIN(G_o) and ADDER, ADDER and JUNCTION, JUNCTION and GAIN(G_p), JUNCTION and DELAY, DELAY and GAIN(a_0) and GAIN(a_0) and ADDER.

At this point, the components and the interconnections are clear; it now remains to implement them in occam. Considering the Lowpass section already investigated, since it constitutes what seems to be a distinct component in itself, this can be defined as a procedure, as shown in Figure 8.6. Note the use of procedures `GainG0` and `Gaina0`, which are simply copies of the archetypal `Gain` procedure, modified to reflect their different coefficient values.

The seven channels are clearly evident, with two of these declared in the `Lowpass` procedure heading and the other five as channels internal to the Lowpass component. Five components compose the Lowpass component and are executed in parallel. By judicious choice of channel names, the occam code can be followed without too much difficulty.

By pursuing the approach outlined, the complete filter diagram structure can be modelled in occam with minimal alterations, typically to increase or decrease the number of output lines in JUNCTION, for example. Assuming that the filter shown in Figure 8.1 is implemented as a procedure, the filterbank is simply defined as the concurrent execution of 128 such procedures, linked in series, as highlighted in Figure 8.7.

```
PROC Lowpass (CHAN OF REAL32 input, output)
  -- Lowpass specific PROCs eg GAINGO
  CHAN OF REAL32 gain G0.add,
                 add.junction,
                 junction.delay,
                 delay.gaina0
                 gaina0.add:

PAR
  GainG0 (input, gainG0.add)
  Adder (gainG0.add, gaina0.add, add.junction)
  Junction (add.junction, juction.delay, output)
  Delay (junction.delay, delay.gaina0)
  Gaina0 (delay.gaina0, gaina0.add):
```

Figure 8.6

```
CHAN OF REAL32 line[Number.Of.Filters]:
PAR
  Startprocess (line[0])
  PAR i = [0 FOR Number.Of.Filters]
  Filter (line[i], line [i+1])
  Endprocess (line[Number.Of.Filters])
```

Figure 8.7

At this stage in the development process, the simple rules underpinning the software components and their interconnection are unable to assist any further. Realization of the concurrent program structure, though significant, only provides the framework within which the subtleties and nuances of the application exist. Consequently, the procedures developed may have to undergo amendment, such as in the case of Figure 8.1, to allow the coefficients belonging to GAIN components to be loaded into the model and passed along its length, so that individual filters can collect their coefficients prior to actual processing of values. Other matters such as deadlock may also require attention if any other input and output is introduced, such as above for coefficient loading.

8.6 Overall system

Whilst fundamentally important, the actual BM model can only be a part of the overall diagnostic system. As shown in Figure 8.8, the model requires extensive servicing and figures as just another concurrent subsystem.

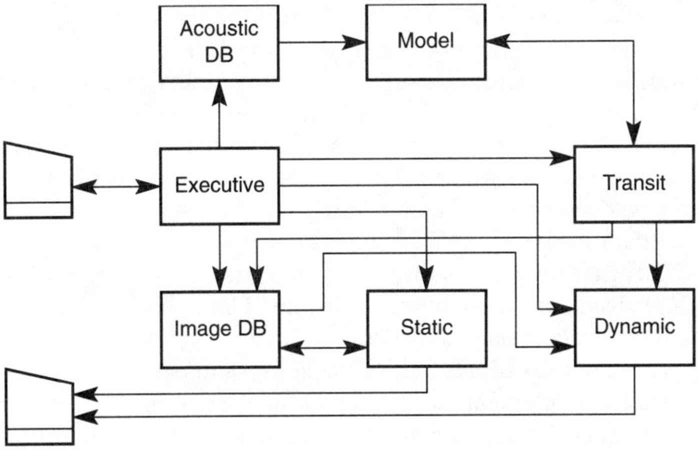

Figure 8.8

Each subsystem implements a definite function, as outlined below:

- ACOUSTIC database – maintains and collects historical and live acoustic data, respectively, for supply to the physiological model;
- MODEL – represents the physiological model, in this case the BM;
- EXECUTIVE – governs and coordinates the execution of the system;
- TRANSIT – manages the retrieval of data from the model, and acts as an intermediate repository for the data on its journey to the dynamic imaging subsystem or the image database;
- DYNAMIC – processes image data representing the action of the model in response to user requirements, either as the data is generated by the model or from historical images retrieved from the IMAGE Database;
- STATIC – facilitates the construction of high quality images of the physiology of the model for use in training and in placing the dynamic images in perspective;
- IMAGE Database – maintains image data representing both the action of the BM and detailed physiological pictures.

By way of clarifying the general relationship of these subsystems to each other, it is illustrative to consider one execution of the model. To run the model, acoustic data must first be selected. This can be historical data retrieved from a fixed disk or live data samples from an analogue to digital converter. The ACOUSTIC DB subsystem is responsible for this, and ensures that the appropriate data is supplied to the model. Before the model is permitted to process acoustic data, the manner in which data, is to be collected from the model must be specified. With respect to the BM this can mean one of two possibilities: either the observation of BM acting along its entire length in response to a specified piece or series of acoustic data, or alternatively the deflection of a specific section of the BM to all of the supplied acoustic data. Once selected, the chosen option is conveyed to the TRANSIT subsystem, whereupon the necessary preparations for data collection are initiated. With model input and output parameters established, it remains only to affirm whether the data in the TRANSIT subsystem is to be communicated to the IMAGE DB for storage and subsequent analysis, or is to be passed to the DYNAMIC subsystem for immediate display. Once this latter decision has been made model execution commences.

The particular function of all of the subsystems, their internal structure and their detailed implementation is beyond the scope of this chapter. Much more pertinent is how these concurrent subsystems are coordinated and controlled. From the above outline of processing it is evident that coordination and control is central and crucial to achieving organized multiprocessing. This key role is performed by the EXECUTIVE subsystem.

8.6.1 Coordination and control

Control is a necessary requirement in any system, but in a concurrent system exercising this requires the cooperation of distinct and disjoint subsystems which are largely autonomous. Achieving the overall purpose of the system, typically vested in some controlling subsystem, consequently reduces to taking full advantage of the cooperative nature of subsystems by effective and appropriately directed coordination of their collective behaviour, to realize a perfectly synchronized effect. In the approach chosen for developing the BM model, this results in control ultimately devolving to a single process, with coordination being effected via channels.

The process within the EXECUTIVE subsystem upon which absolute authority is conferred is known as the **runcontroller**. This process, directly or indirectly, causes the other subsystems to interact in the order outlined above. How this is achieved differs according to the function of the respective subsystem, and for this reason cannot be comprehensively dealt with within the confines of this chapter. Nevertheless, a valuable insight can be gleaned by looking at the interaction between the runcontroller and those subsystems directly connected to the model, namely ACOUSTIC DB and TRANSIT. These two subsystems, in addition to the EXECUTIVE subsystem, form the environment of the model, and are therefore particularly important.

8.6.2 Acoustic database

This subsystem performs two main functions from the perspective of the model. Although it primarily supplies acoustic data to the model as its major function, whenever the overall system is initially loaded it also supplies the coefficients that endow the model with its physiological properties, as is typical of such models. Since these coefficients are crucial to the quality of the output and an integral aspect of the model, it is essential that their positioning within the model precede processing of actual acoustical data.

This latter function of the two outlined above is implemented by a coefficient control process which immediately upon system loading attempts to feed coefficients to the model. Since there are ten different coefficients per filter, 1280 coefficients, each represented by a 32-bit real number, must be communicated to the model. To minimize the effect of this 'one-off' operation upon the concurrent architecture of the model, the coefficients are dispersed throughout the model using the channels defined for processing acoustic data. Whilst this results in some additional code being initially executed in some of the processes which implement a filter, the internal distribution and progression of coefficients within and between filters in the filterbank can be achieved without disturbing the general process and channel structure of the system. Thus the coefficient control and filter processes must cooperate in loading the coefficients before the proper processing of acoustic data can commence. To ensure that this occurs the runcontroller takes no action until the coefficient control process signals the completion of its activity, as indicated in Figure 8.9. Following this signal, the coefficient control process terminates and the filter processes enter their predominant mode of operation, namely the processing of acoustic data.

Once the runcontroller has established that the model is primed with coefficients it solicits the source of acoustic data from the user. There are four such sources in two categories, based on whether the data is generated or not. The two sources of generated data are Impulse/Response and Sine, which are useful for accuracy assessment of the model. The other sources produce actual acoustic data, and can either be historical or live. In the former case, the data is retrieved from the acoustic database, and from the latter it is collected directly from the external environment.

Each of these data sources is represented by a process and executes concurrently, allowing the production of data to take place at the same time as data supplied previously is being processed. To divorce the model from the intricacies of the data supply a single stream of data is delivered to it, as illustrated in Figure 8.9. Besides greatly simplifying the interface to the model, this permits the inclusion of other data sources in the system with no subsequent impact on the model.

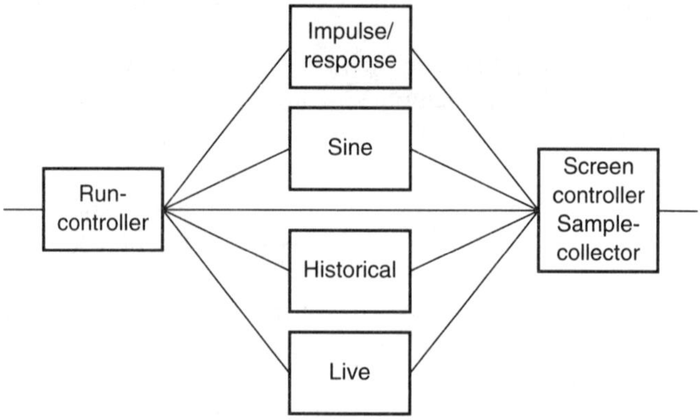

Figure 8.9

Figure 8.9 highlights how the choice of data source is implemented in occam 2. First the runcontroller signals the appropriate data source, effectively switching it on. A data source reacts to this by immediately supplying data along its respective output channel. To funnel the acoustic data into the model along one channel, a **samplecollector** process multiplexes the four incoming channels. Thus although all of the processes are concurrent, only one of the data sources can supply data at any point in time. To ensure that all data sources are always ready to be activated, even after supplying data, they each execute within an infinite interaction which begins with an input command connected to the runcontroller, followed by the body of code which supplies the data. Thus a data source, having delivered its data, returns to the start of the interaction to await reactivation by the runcontroller. With data sources structured in this fashion, they each enforce mutual exclusion upon each other with respect to the production of data. To prevent the model from processing the selected data before the mode of data collection has been established, the samplecollector is prevented by the runcontroller from passing on the data. This is achieved by forcing the samplecollector to suspend its activity until it receives a communication from the runcontroller.

The multiplexing evident in this part of the system is a consequence of the synchronous point-to-point communication scheme used in occam and employed by the transputer. This approach prohibits processes from sharing a single channel. Although this might appear very restrictive at first, in practice it is not difficult to implement, as shown in Figure 8.10, and it leads to a process and channel structure which can be altered in a straightforward manner. For example, a data source can be added independently of other data sources, requiring only the definition of two new channels and minimal amendments to the runcontroller and samplecollector processes.

A further important consideration is that when processes and channels are mapped to transputers and physical links it is a relatively simple task, compared to the situation that would prevail with shared channels. With shared channels permitted, besides the problem of detecting them, the placement of

```
PROC Samplecollector (CHAN OF ANY RunControllink,
                      []CHAN OF REAL32 Source,
                      CHAN OF REAL32 Sink)

  WHILE TRUE
  CHAR any:
  REAL32 data:
  SEQ
    RunControllink ? any
    ALT a.source = [0 for 4]
    source [a.source] ? data
    SEQ
      Sink ! Data
      SEQ i = [0 FOR NumberOfSamples - 1]
        Source[a.source] ? data
        Sink ! data
```

Figure 8.10

processes on transputers would be affected by the need to ensure that the collections of processes at either end of a shared channel were located on the same transputer. Depending upon the particular processes sharing a channel, this could prove rather awkward, certainly tedious and quite error prone.

8.6.3 TRANSIT

The key function of the TRANSIT subsystem is the organization of the retrieval of data from the model in accordance with the mode of observation chosen. In essence, this reduces to data being streamed out of all 128 filter processes for a specific acoustic sample, or all of the data from one nominated filter process. As shown in Figure 8.11, this is achieved by a **datacontroller** process which, in response to the provision of the respective information by the runcontroller process, coordinates each filter appropriately.

Since there are 128 filters, each potentially producing data at the same time, the TRANSIT subsystem implements a **databuffer** process on a one-to-one ratio with filter processes. Consequently, each databuffer process must be individually informed of the mode of observation that applies to its respective filter process. If this were not done, communications deadlock would rapidly develop. Although it might appear from this that synchronous communication presents more of a disadvantage than an advantage, this is not the case. In fact, this very scheme allows the runcontroller process to immediately embark on the next stage in initiating model execution as soon as its initial communication with the TRANSIT subsystem has completed. Even though the TRANSIT subsystem has data retrieval as outlined above, the runcontroller can initiate model execution while the TRANSIT subsystem concurrently proceeds to complete its 'setup' task. This flexibility is a direct result of the properties of synchronous communication.

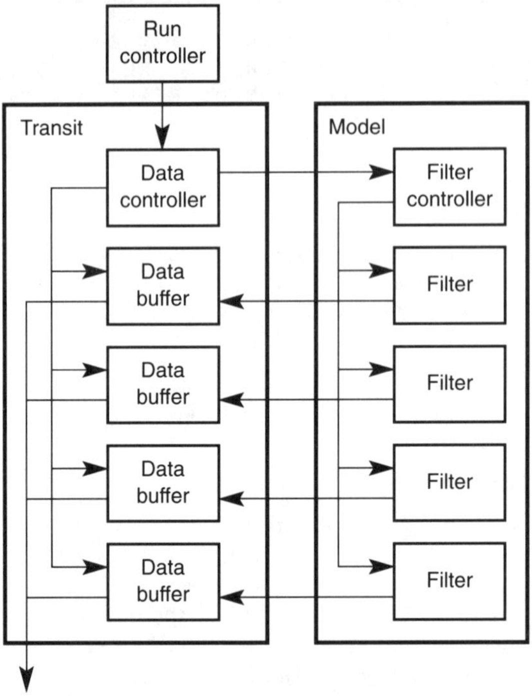

Figure 8.11

Once model execution has begun, data flowing out of the model is lodged in the databuffer processes within the TRANSIT subsystem. These processes, besides their primary function of unloading data from the filter processes and conveying it to the DYNAMIC subsystem, perform another valuable role. This role addresses the need to prevent, or at least minimize, degradation of model execution through data not being removed fast enough from the filter processes. This explains why there is a concurrent series of databuffer processes. In an attempt to further minimize the likelihood of degradation occurring, each databuffer process contains a cyclic buffer to improve the ability of the TRANSIT subsystem to equalize the speed of operation of the MODEL and DYNAMIC subsystems.

8.7 Implementation evaluation

8.7.1 The Basilar Membrane model testbed

This section details the operation of the filter program and test harness developed for performance evaluation. At the outermost level the only inputs and outputs to the testbed are by way of the screen and keyboard channels.

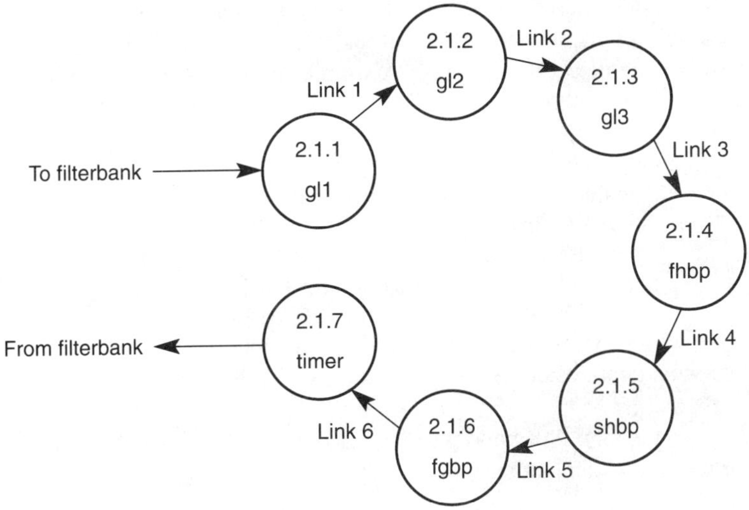

Figure 8.12 Level 2 diagram.

It should be noted that the filter program used here would under normal circumstances be replicated on a network of transputers, and each filter would be supplied with its own set of coefficients from an external file at runtime. For the purposes of performance evaluation, the coefficients were simply hardcoded.

Figure 8.12 illustrates the actual filter process. It has one input which consists of a number of REAL32 samples and a single REAL32 output, which represents the time taken to perform the operations involved.

The gl1 process simply reads in a value from its input channel, multiplies it by the constant oneminusa0, then outputs this result on its output channel.

The gl2 process reads in a value from its input channel, adds a0 multiplied by its previous output value, then outputs the result. On the first call of this process the last output value is taken to be 0. The fact that it needs to store its last output value makes it difficult to improve its performance. Its operation can be summarized as:

```
set last output value to 0
read invalue from channel in
for z := 0 to number of samples – 1 do
        set outputvalue to invalue + a0 * last outputvalue
        transmit outputvalue to next process on channel out
        get next inputvalue from channel in
output invalue + a0 * last outputvalue on channel out
```

Process gl3 simply takes an input from a channel, applies the transform:

```
outvalue := invalue * GpxKg,
```

then outputs the result on the output channel.

Process fhbp (first half band pass) is the first process to have a two stage delay or feedback mechanism. The output value from one sample is used in the processing of the next two samples. A simple form of queue is implemented to let values filter out. The action of this process can be summarized as:

```
read an input value from the in channel
for z := 1 to the total number of samples do
        set Vp1 to (input value + (b1 * y1)) – (b2 * y2)
        set y2 to the value of y1
        set y1 to the value of Vp1 (for next iteration of the loop)
        transmit Vp1 on the out channel
        read a new input value from the in channel
transmit (input value + (b1 * y1)) – (b2 * y2) on the out channel
```

Process shbp (second half band pass) is the other process which has a two stage delay or feedback mechanism:

```
read an input from the in channel
for z := 1 to the total number of samples do
        set Vp2 to (input value – (a1 * y1)) – (a2 * y2)
        set y2 to the value of y1
        set y1 to the value of the last input value (for next iteration)
        transmit Vp1 on the out channel
        read a new input value from the in channel
transmit (input value – (a1 * y1) – (a2 * y2) on the out channel
```

fghp (final gain, band pass) is another relatively simple process. It takes an input value, multiplies it by the constant Gz then outputs the result.

The network timer process is responsible for accurately measuring the time required to complete the processing. It should be noted that it does not have to read in values until the pipeline is saturated, as it is placed at the end of the pipeline. This means that the first result received will indicate that the pipeline is full. Its operation is as follows:

```
read in enough values until the pipeline is saturated
read the timer channel, place result in start
read in a further 2048 samples
calculate the elapsed time
transmit elapsed time back to host
```

8.7.2 General performance evaluation

As a basis for comparison some of the industry standard benchmarks were implemented; however, benchmarks can only make generalizations about the performance of a machine. The acid test is how it performs on real life problems.

Table 8.1. Basilar Membrane model performance.

Machine	*Operating System*	*Language*	*Time(s)*
DEC VAX 8700	VMS 5.3	Pascal	4.70
DEC VAX 6310	VMS 5.3	Pascal	7.63
DEC VAX 11/785	VMS 5.3	Pascal	15.78
DEC VAX 11/750	BSD 4.2 UNIX	Pascal	238.68
Micro Vax 6300	Ultrix	Pascal	94.67
SUN Sparc1+	SUN OS	Pascal	5.87
SUN 3/160	SUN OS	Pascal	272.40
IBM 3090 180s	VM/370	Pascal	0.7919
IBM PS/2 mod 70	MS-DOS	Pascal	117.60
IBM PS/2 mod 60	MS-DOS	Pascal	259.69
OPUS PC V	MS-DOS	Pascal	270.06
OPUS PCIII + 287	MS-DOS	Pascal	155.38
IMS T800-17	N/A	Pascal	10.928
IMS T800-17	N/A	3L C	10.235
IMS T800-25	N/A	occam	0.322
6 × IMS T800-25	N/A	occam	0.176

The complete 128 filterbank Basilar Membrane model was implemented on a selection of hardware, and Table 8.1 shows the initial results for the model without any performance tuning attempts whatsoever. On the multi-user machines such as the VAXs and the IBM 3090, attempts were made to perform the test with a limited workload.

Maximizing the use of on-chip cache memory

Transputers provide a very fast cache memory of either 2 K or 4 K, which is located at the start of memory (address #80000000). The cache has a typical access time of between 50 and 66 ns, depending upon which transputer is being used. Off-chip memory, which makes up the rest of the 2 Mbytes on the Transtech TMB08, has an access time of about 120 ns. This degradation by a factor of two can make a significant difference to the performance of memory intensive programming tasks. Clearly, if variables can be placed in the cache, a significant speed-up factor can be attained. Occam provides two constructs to enable variables to be mapped to a specific hardware address. Before starting, it should be noted that workspace is used to refer to the data requirements of the current process. WORKSPACE is used to refer to the actual area of the cache set aside to store the contents of the workspaces.

The 'PLACE<var>AT' construct places an individual variable at a memory location, but causes problems because the variable will be outside the local workspace. When occam tries to access any variable, it takes an offset from the W register to find the actual address of the data. To access a variable placed in this way, the offset has to be calculated from a fixed base held in the A register. The extra code needed to swap the base address between the W and the A registers will actually cause a degradation in performance. The swapping may be achieved using the occam abbreviation function VAL, or by resorting to assembly. The abbreviation is a very powerful feature which can be used to bypass a lot of occam's integrity, security and scope restrictions to improve performance.

A better solution is to allow the entire workspace to be placed in the WORKSPACE area of the cache memory. The occam compiler will do this by default for individual variables, but arrays will be placed in off-chip RAM. If the array is heavily used, or if a time-critical task is performed on it, it is also possible to place it in the WORKSPACE area. This is effected by use of the 'PLACE vector IN WORKSPACE' construct.

Unfortunately, the individual processes of the Basilar Membrane model are so small that both the data and the execution code fit onto the on-chip cache. No improvement is possible. This fact did, however, create problems for other stages of the work, as they increased the code size, which then meant the code could not reside on the cache. This resulted in poorer performance figures. If the original code had been placed in standard memory, the modifications would have yielded reasonable performance increases.

Opening loops

It was noted during an investigation of assembly language that the actual model code was only about 400 bytes, and the data requirement was only another 30 bytes or so. The T800s on which the performance work was being carried out had a 4 K on-chip cache. This meant there was wasted space on the cache. Opening loops is one way in which to use this space efficiently.

The general principle is that you perform the same operation several times inside the loop, but reduce the number of times the loop is performed by the factor of the number of times the operation is performed. This means that the delays introduced by the loop checking code and the actual jump instruction are reduced by the number of times the operations are repeated.

The experiments carried out into performance optimization using loop expansion also showed how the use of the cache memory can affect the speed of execution. As the number of loop expansions increases, the size of the code grows. Eventually, a point is reached where the performance starts to suddenly decrease, instead of increasing. This is the point at which the code no longer can fit into the cache, and so has to be placed in the standard, slow RAM (Figure 8.13).

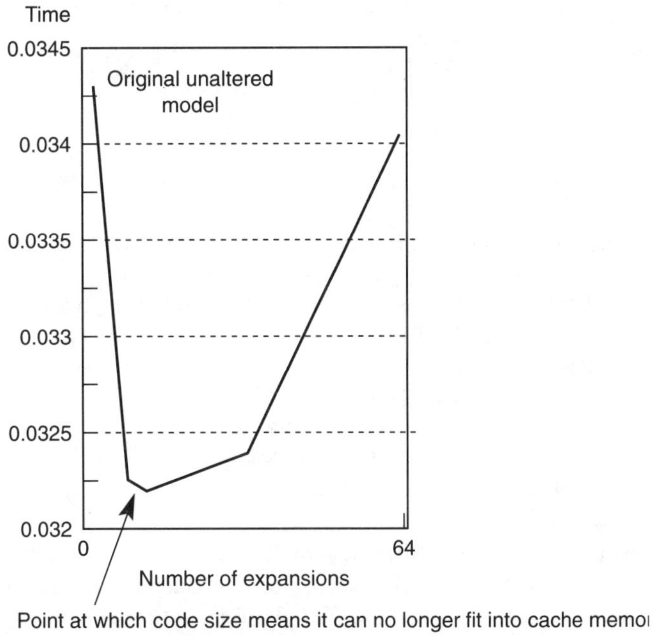

Figure 8.13 Results of loop expansion on model running on one T800.

Maximizing link performance

For many applications, the relatively slow speed of the links is the weakness in the transputers' chaining of processes. Any way to improve link performance will improve the performance of the overall system.

The links are autonomous DMA engines, capable of transferring data bidirectionally at rates up to 20 Mbits/second. There are two basic methods which can be used to improve link performance: decoupling and large link transfers.

Decoupling Since the links provide occam processes with synchronization, a situation can arise where the process is sitting idle, waiting to send or receive a message from another processor. Decoupling provides a means by which to free the processor to carry out further work. Decoupling basically adds a buffer process with the I/O requirement. It should run in parallel with the current process. Using the buffer, the message is sent to the buffer and held until the other processor is ready to receive it. While the buffer process is trying to transmit the message, the main process can carry on. Priority may be assigned to either the buffer process or main process, depending upon which is more critical. For example:

```
PROC bucket (CHAN OF INT in,out)
  WHILE TRUE
    SEQ
      in ? data
      calculation
      out ! data
:
```

This code can be improved considerably by rewriting as:

```
PROC bucket (CHAN OF INT in,out)
  WHILE TRUE
    PAR
      buffer (in,a)
      calculation (a,b)
      buffer (b,out)
:
```

where `a` and `b` are buffer channels and `buffer` is the buffer process.

The effect of decoupling on the Basilar Membrane model was tested in steps (6) and (7) of the parallel pipeline tests (see below).

Large link transfers The time taken to initiate a message transfer on a link is much greater than the time required to actually transmit the bits. It follows, then, that messages should be made as long as possible. It is possible to 'batch up' transfers on links. As much use should be made of this capability as possible. For example, instead of using:

```
SEQ i = 0 FOR arraysize
  out ! array.data[i]
```

use:

```
  out ! [array.data FROM 0 FOR 300]
```

which requires only one initialization as opposed to the 300 of the previous version.

The potential for performance improvement on the Basilar Membrane model was tested during steps (2) and (3) of the parallel pipeline tests (see below).

8.7.3 Single filter performance analysis/optimization

The goal for this stage of the work was to produce a real-time response from a single filter running on a network of transputers. The general performance enhancements from the previous section were incorporated into an initial test configuration, which was basically a straight, single pipeline consisting of six pipe stages. This code was used as a basis for all the work in this section.

There were five separate timing methods employed, which could each produce a slightly different result. In some cases, the timing was taken on the controller program, an EXE running on the host, which was subject to the interference from host I/O, and from the fact that the receiving and timing code may or may not have been located in the cache RAM. The timers which ran on the network also had problems associated with them. To get an optimum timing from the pipeline, the pipeline must be full. Attempts were made to take this fact into consideration on the timing processes by using code similar to the following:

```
SEQ i = FOR no.of.samples
  SEQ
    IF
      i = pipe.length
          clock ? start
      TRUE
        SKIP
    in ? dummyvalue
    clock ? stop
```

This code did not, however, give a true measure of the pipeline's speed, as the time taken for the IF statement to execute, at each iteration of the loop, is also included.

With a knowledge of TMB12 hardware limitations, and some idea of the number of processors which would be needed at each stage of the work, it was decided to produce two standard timers. One was an EXE which was used only for debugging purposes and cases where there were no free transputers available on the network. The other was a process which could be incorporated into a PROGRAM, and which was more accurate than the EXE version. The PROGRAM version read a start value from the timer, then read 2048 output values. A stop value was then read from the timer, the elapsed time calculated and the result passed back to the host. A version was also written which used the high priority timer. This simply increased the resolution of the timer by a factor of 64, that is, differences of up to 64 microseconds could be measured, instead of just one microsecond.

A decision was also made to take timings only from PROGRAMS which were compiled in occam's REDUCED mode. This mode means that checks for the error condition being flagged are not made. The checking code can be very time consuming. For example, the Savage Benchmark will run on a single T414 in 6 seconds, when compiled in REDUCED mode. In HALT mode, the default option on the compiler, the same code takes about 26 seconds to complete. REDUCED mode cannot be used on an EXE, which uses input or output to the host, as Inmos do not supply Input/Output libraries for REDUCED mode. This is another reason for producing the timer as a process in the PROGRAM, and having a source process feed the pipeline.

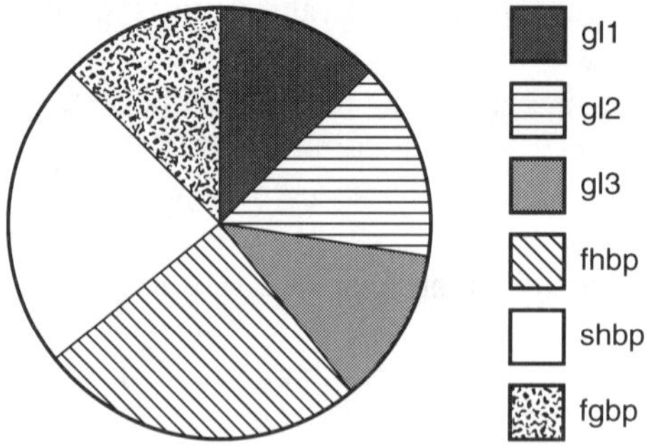

Figure 8.14 Results of pipeline analysis on the six stage pipeline. Note that fhbp and shbp account for almost 50% of all processing. These results take communication overheads into consideration.

Working in either HALT or STOP mode also means that timings are biased in favour of smaller processes in each of the pipeline stages, as they will permit a greater throughput of results before the error check code is executed.

Step 1: Pipeline analysis

Using the six stage sequential pipelined filter as a base, basic performance metrics were calculated for each stage. This was achieved by simply replicating each of the six basic stages across a network of 16 processors. A series of 2048 values were fed in at the pipe head, and a timing taken for the pipeline to process the results. The timings were taken only when the pipeline was full.

This work enabled the bottlenecks in the system to be identified for further performance optimization work to be carried out, and also identified which of the processes could be combined to optimize for cost. The results of this work are given in Figure 8.14.

Step 2: Link analysis

Realizing that the other limiting factor on the six stage sequential pipeline was the operation of the interprocessor links, a selection of test programs were written which enabled the performance of the links to be analysed.

The first test used the standard filter program, but the test was carried out on the machines at the Northern Ireland Regional Transputer Centre, as the transputers there had their link speed permanently set to 10 Mbits/second. All other variables such as the processor type, its operational speed and the speed of its associated memory were kept constant. The result of this test showed that when the link speed was halved, the overall execution time increased by about 25%. This suggested that the link operation was critical in the Basilar Membrane model, and opened the way forward for further investigation into link performance.

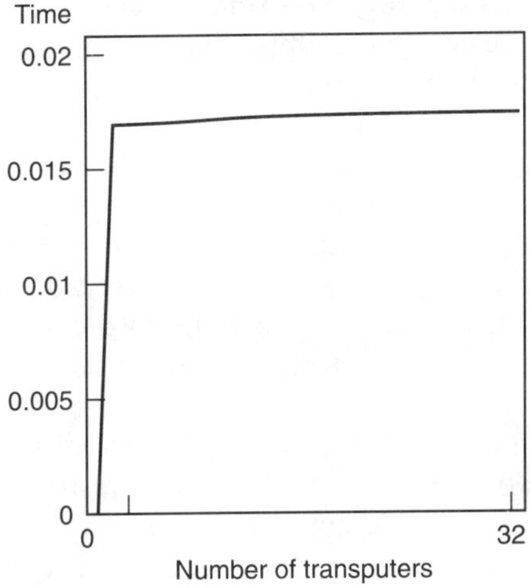

Figure 8.15 Link performance results. All links were set at 20 Mbits/second. Times are for the transmission of 2048 samples.

The further investigation took the form of a series of test programs which consisted of **buckets** replicated a number of times on a hardware pipeline. A bucket in occam can be thought of as a vessel which simply carries a value from one end of the pipe to the other, with no intervening processing (also known as **passthru** stages).

The bucket test was carried out using 1, 2, 15 and 32 transputers. The result in each case was the same: 0.016 seconds (see Figure 8.15). This figure (0.016 seconds) had no processing involved other than a straight assignment from input variable to output variable.

The parallel pipelines

The Basilar Membrane model has a pair of two stage delays and one single stage delay, which complicate the process of placing two or more of the model pipelines in parallel. The delays are variables which hold a value that was used in processing the last input sample, or the input sample before that.

Further complications were introduced by the use of the TMB12 boards, as the transputers on these boards are hardwired in a pipeline network, from link2 to linkl. This meant that any communication between the parallel pipelines would be restricted to using the two remaining links on each transputer. Some method had to be found to place the separate model pipelines on the actual hardware pipeline on the TMB12. The TMB12 also presented a problem in that the first transputer in its pipeline could not be used to split the input to the two splits, as it had only one input link and one output link available. Its other links were permanently connected to the Network Controller (a T212) and the Reset/Analyse lines of all the other transputers on the board.

Step 1 The first step of this stage was simply a check to make sure that replicating the model's pipeline and splitting the input would, or could, actually result in a performance increase.

A simple two-way split was designed (Figure 8.16), which basically replicated the six stage standard model pipeline twice, and split the number of input values, between the two splits. Synchronization between the input values and the output values was maintained, but this version took no account of the three feedback (delay) stages. The actual feedback processing was carried out to reflect the true processing load, but it did so only on every other true input sample value from the host. In effect, what was produced was two totally separate filters, each fed by a single source, and combining their outputs for a final timing of all 2048 samples. The object was not to produce an accurate implementation of the model, but was just to test that a two way split could input, process and output 2048 samples in less than 0.01 seconds. It succeeded. The configuration of the network used for this test is given in Figure 8.17.

Step 2 This step took the timing code used in Step (1) and introduced true feedback loops between the stages, as shown in Figure 8.18. The aim of this step was to produce an accurate implementation of the BM model. To do this, communication was needed between all the stages of the pipeline which had a feedback system of any sort. The communication carried the feedback values needed from the previous sample's processing at the same stage of the model's pipeline. This would always have been on the other split. All communication between the splits was bidirectional. In total, three of the processes needed some form of communication to their counterpart across the split. Two of these three had to send two values across the split. In this case, the two values were simply sent in sequence along the same link. Although spare links were available (Figure 8.19), this practice should have been quicker than having two separate links and sending each value in parallel, as time would not be wasted initiating channel communication (see the point made on Large Link Transfers – p. 216).

Step 3 Step 3 extended the step (2) implementation to test the Large Link Transfer theory. This was achieved by allocating both of the spare links on each transputer, which had a process that needed a feedback variable, to a communication channel (shown in Figure 8.20). The two processes (fhbp and shbp) which required feedback from both the previous samples transmitted both the feedback variables required in parallel, after that process had completed. This meant that at the end of each execution loop these processes had to initiate three channel communication transfers. A reasonably large degradation in performance was expected.

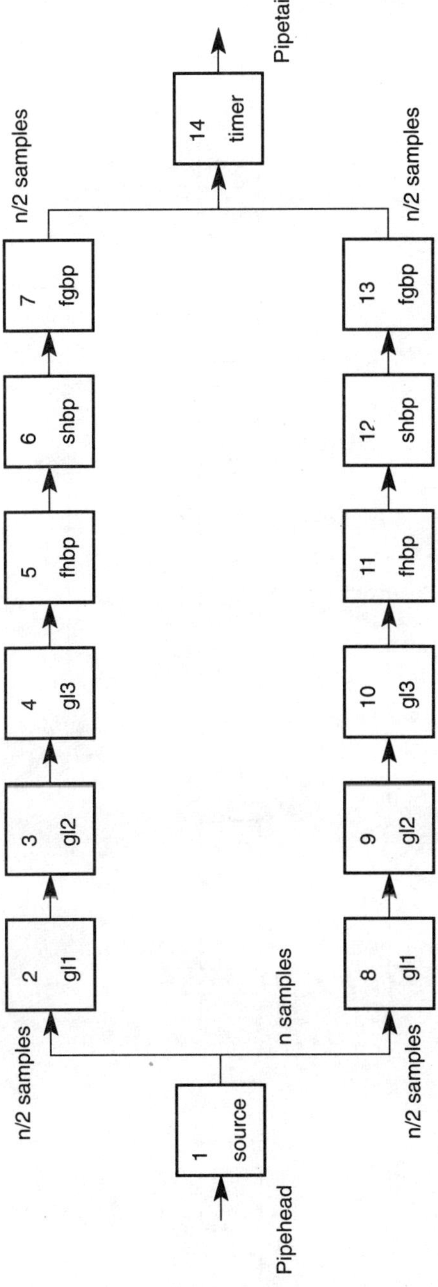

Figure 8.16 Logical representation of Step (1).

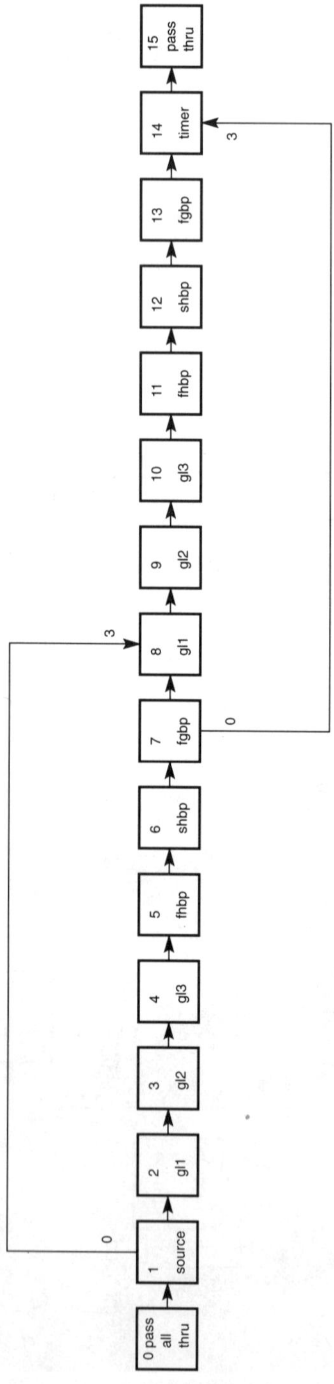

Figure 8.17 Network configuration for pipeline used in Step (1).

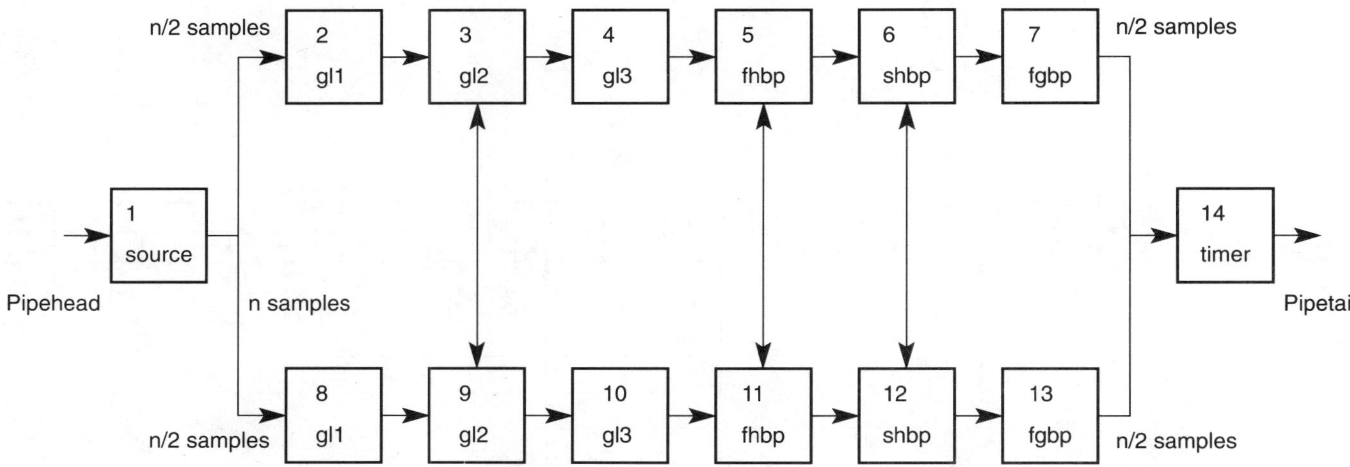

Figure 8.18 Logical representation of Step (2).

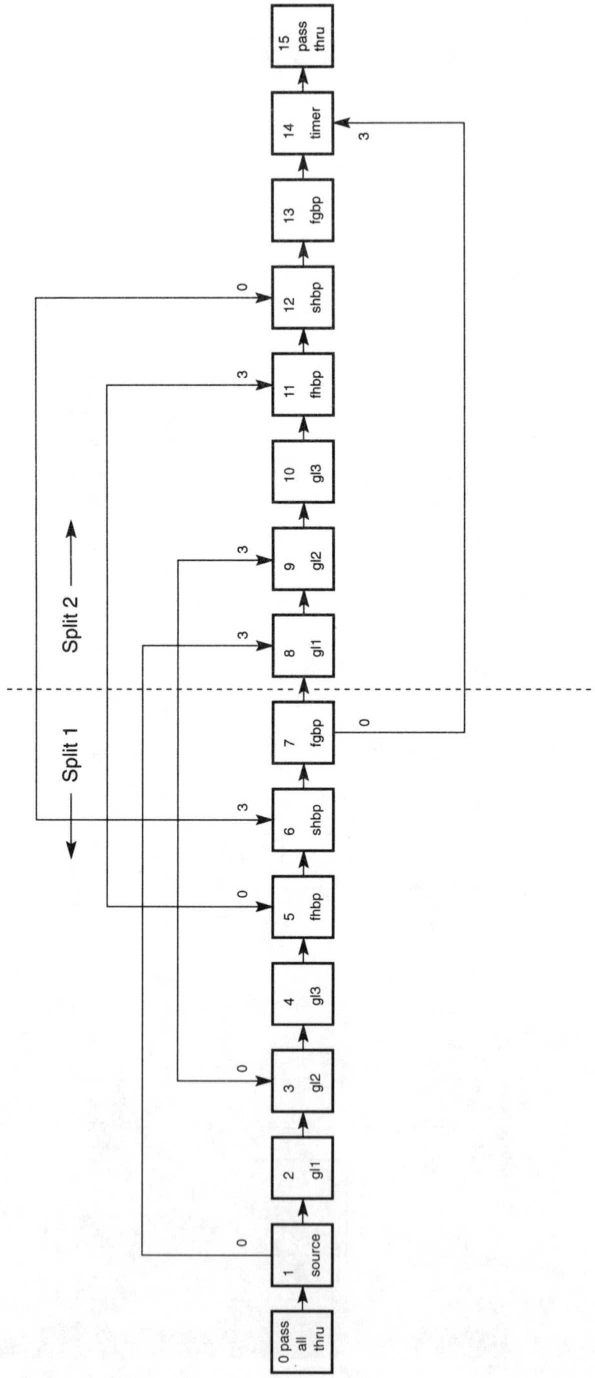

Figure 8.19 Network configuration for pipeline used in Step (2).

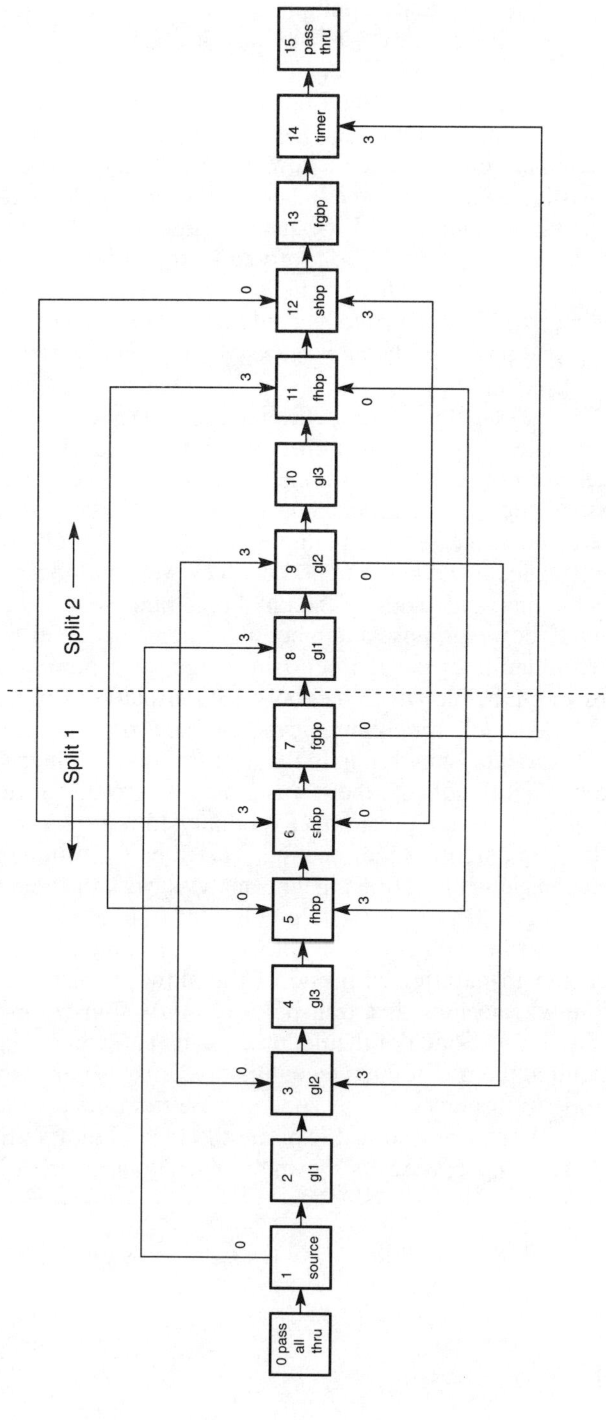

Figure 8.20 Network configuration for pipeline used in Step (3).

Step 4 After examining in detail how the code actually worked, paying special attention to the feedback loops, the configuration of step (3) was used again as a basis for step (4). By this stage, it was obvious that link speed was the major performance limiting factor. In an attempt to reduce the need for communication, the actual code which produced the feedback loops was examined in detail. The processes with the two stage feedback did nothing more than hold a value in a two element queue. When a new value was fed into the queue, the values would shuffle down a level, and the one at the end would be lost. Having just two parallel splits meant that if both queue positions were sent to the other split, the first one sent (yl) would be returned to the current split to fill the queue's second position (y2). Clearly, there was no need to send the second queue position each time, as the process could just move the current feedback element to the second position when it was ready to take in a new sample.

Step 5 Step (5) was an attempt to reduce the time consumed by feeding just two split pipelines. A third split was introduced (Figure 8.21), which meant that the time taken to distribute the samples should have been reduced. Each separate process would then be replicated three times. The feedback interconnections between the processes were simply arranged as a cyclic loop. In the cases where two feedback values were needed, they were simply transmitted in sequence down the link, and taken off the link in the same sequence.

Step (5) did, however, pose a number of implementation difficulties. All the previous work had been carried out using a six stage pipeline replicated twice on a hardware pipeline which had only 16 transputers. The first of these had no spare links available for communication to other processors, so it had to be used simply to pass the samples up to the next transputer, which divided the samples to the two splits. Finally, there was a processor used for recombining and timing the output, meaning that effectively only 13 transputers were available for actually implementing the pipeline. The three split implementations used the knowledge gained from the pipeline analysis work to combine three of the six pipeline stages, leaving a basic pipeline of four stages, which could easily be mapped onto the remaining 13 processors. In practice, another extra process, which ran in parallel with one of the slow processes on the BM model's pipeline was needed, as a transputer has only four communications links, and the TMB12 uses two of them already in its hardwired configuration. One process required the use of three reconfigurable links, so the extra process, on one of the other transputers, simply took a value from the previous processor's hardwired link, and passed it down one the reconfigurable links to the combiner and timer. This topology is shown graphically in Figure 8.22.

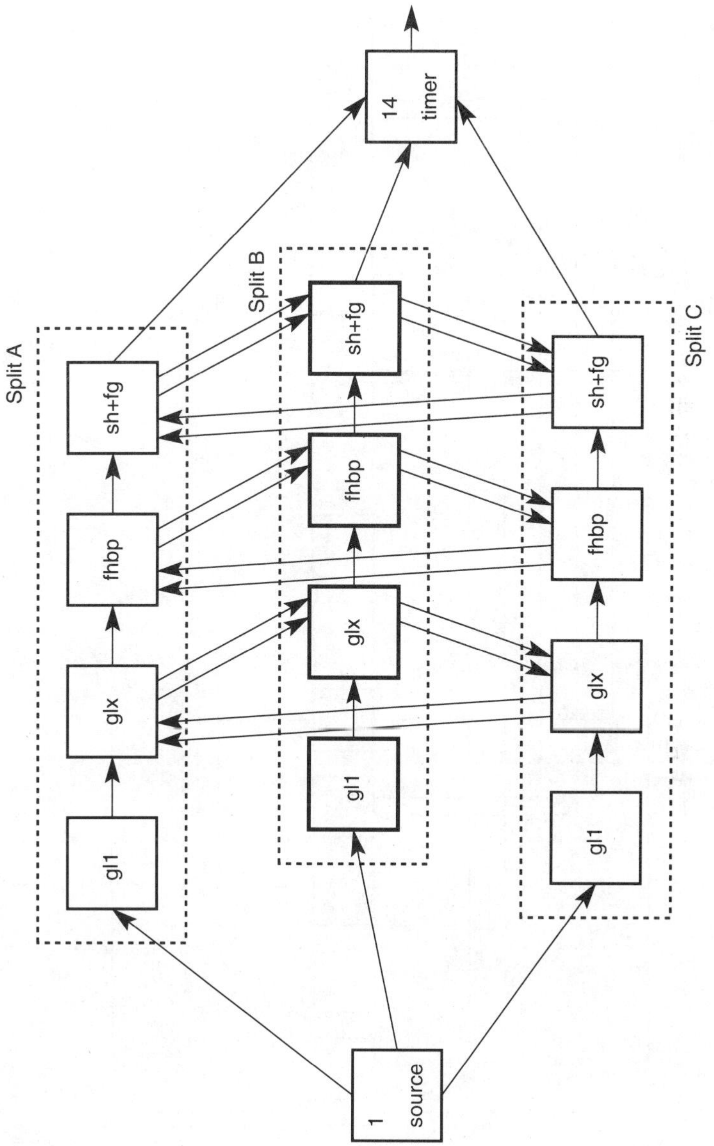

Figure 8.21 Logical representation of Step (5).

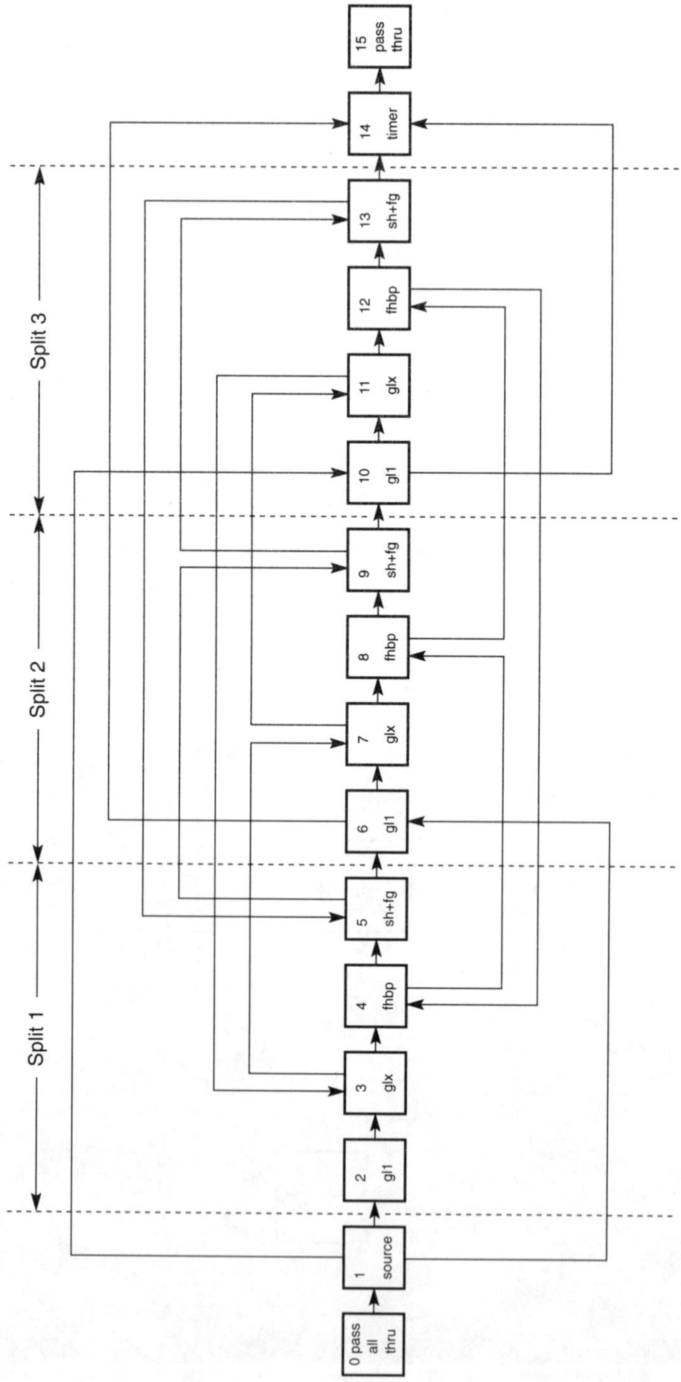

Figure 8.22 Network configuration used for Step (5).

Step 6 Step (6) took the knowledge of the feedback variable 'queues' and applied it to the three split implementation. If, for example, split A was processing the 100th sample, it could transmit its feedback output to split B (processing the 101st sample), which would need it for the first value in its feedback queue (B's second queue value would have come from that associated with the 99th sample's processing). The same feedback output from split A could also be sent to split C (processing the 102nd sample), to be used in the second position of the queue (C's first queue value having come from the 101st sample's feedback output).

This method of operation permitted C to read the feedback value transmitted from A, while B was carrying out its processing. This should have cut down the time wasted due to process synchronization on the channels.

The logical layout, and the actual transputer configuration, of step (6) are shown in Figures 8.23 and 8.24.

Step 7 Step (7) simply took step (6) and applied buffering processes on the channels which supplied values to (Zth + 2) processor. This step surprisingly produced a slower time as the regular nature of the output meant the buffering process was really only effective for the first sample. After that, it simply became a process scheduling overhead.

The results of the parallel pipeline analysis are summarized in Figure 8.25.

The 128 filters

Up until this stage, all performance work had been concentrated on producing a real-time response from a single filter placed on a bank of transputers. This stage of the work was concerned with the performance of all 128 digital filters.

The basic configuration which was used to perform these tests was a single transputer with a number of Basilar Membrane filters pipelined on it. Each of these single transputers was then pipelined together.

The tests were performed on 1, 2, 4, 8, 16 and 32 Inmos T800-25 transputers, all of which are factors of 128. This meant that the load on each transputer would be equal, so no transputer should be a bottleneck on the pipeline. Figure 8.26 highlights the results of this analysis.

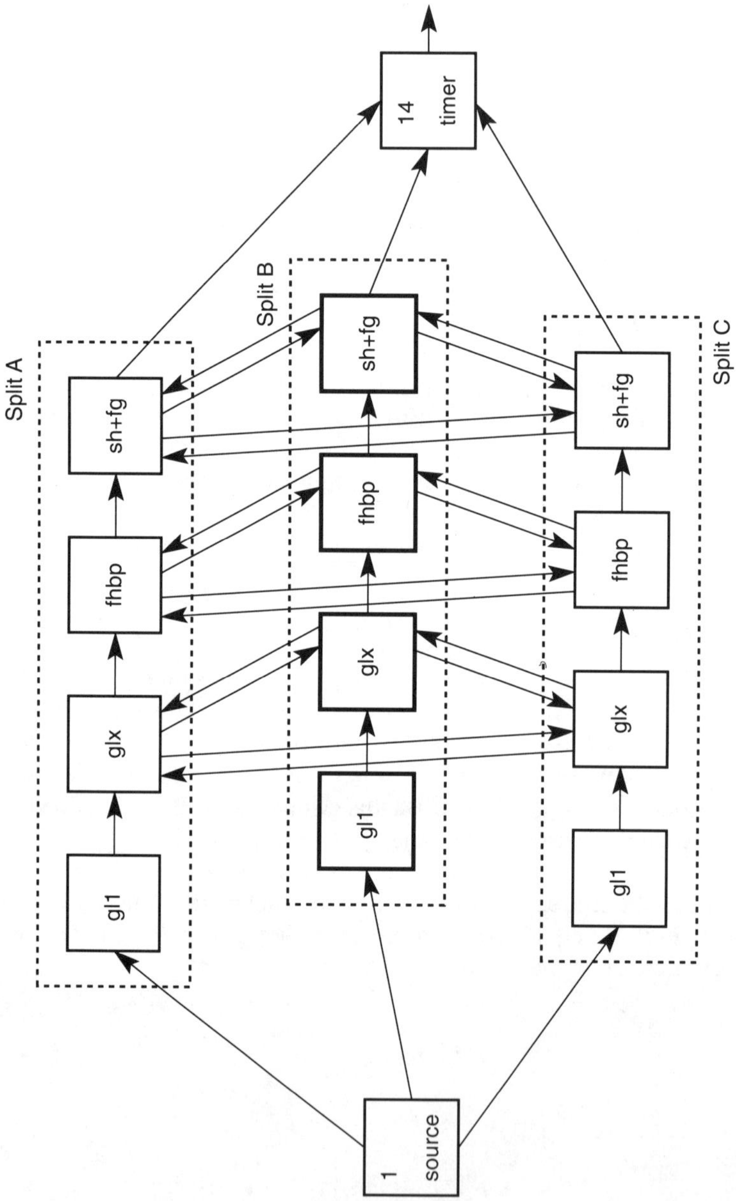

Figure 8.23 Logical representation of Step (6).

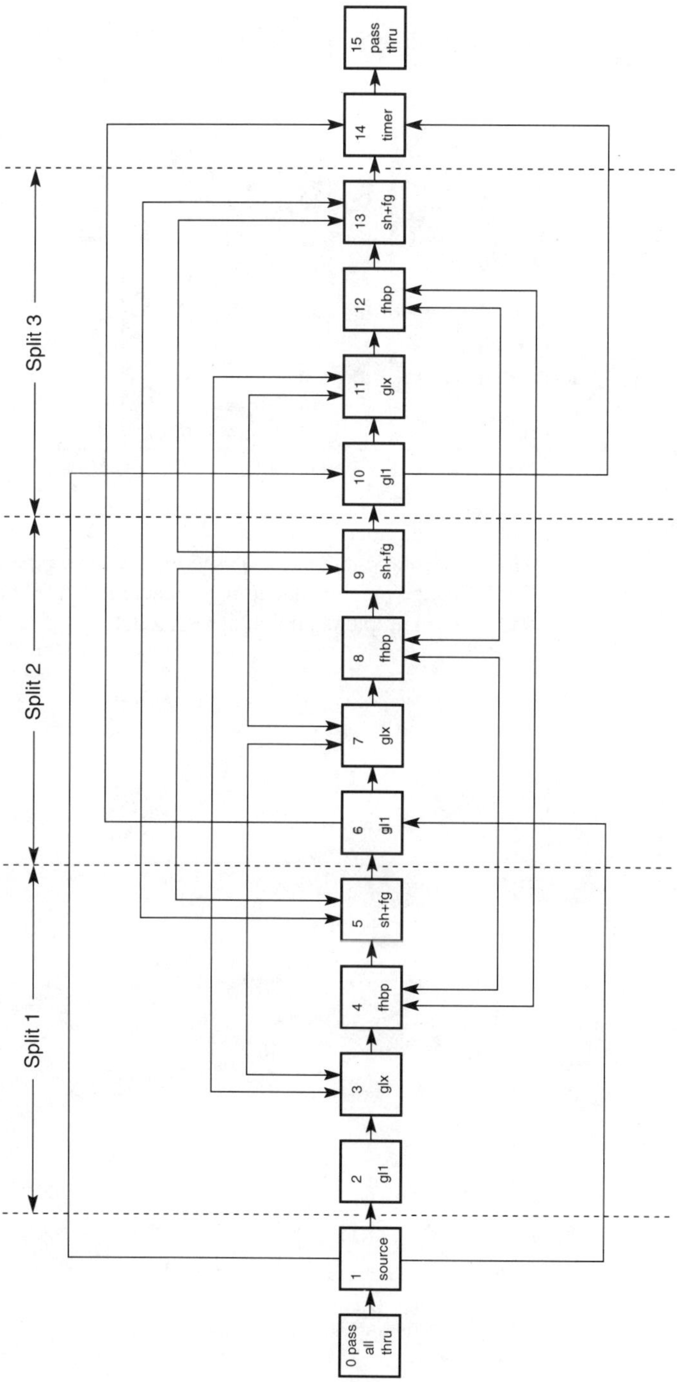

Figure 8.24 Network configuration used for Step (6).

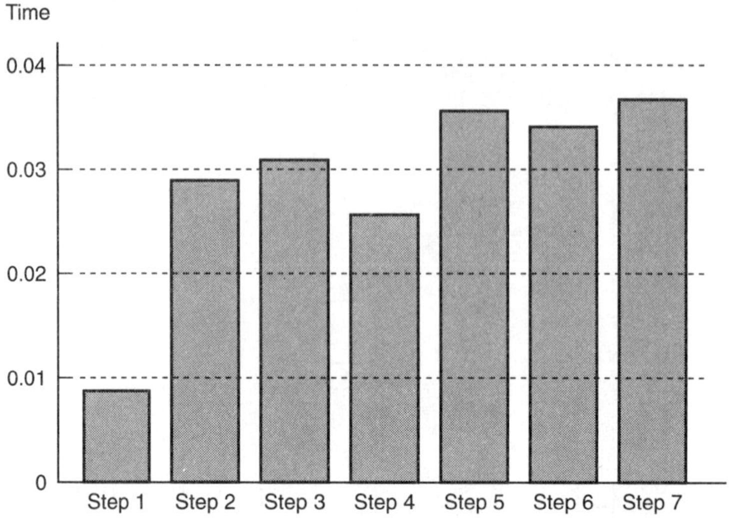

Figure 8.25 Results of the 'parallel pipeline experiments'. All timings performed on T800s with links set at 20 Mbits/second. Steps (1)–(4) used the six stage pipeline as a basis, steps (5)–(7) used the four stage pipeline due to implementation problems on the TMB12. Step (1) was not an accurate implementation of the model.

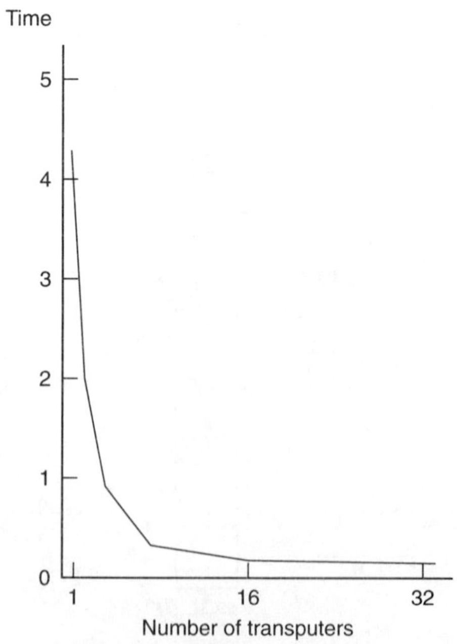

Figure 8.26

8.8 Conclusion

Our objective has been to develop simple rules to allow the development of a concurrent program structure able to model systems expressed as digital filter diagrams. These rules have been aimed at those not familiar with concurrent programming languages, and occam in particular, and who would like to explore the possibility of parallel approaches in their domain of research. From experience with one filter diagram (Figure 8.1) relating to a specific model, we have been able to construct and take to implementation, using the simple rules outlined, an occam program which has proved to be an accurate analogue of the digital filter design.

There are, of course, many more issues involved in reaching a final implementation, often requiring techniques to improve performance such as process collapsing and the mapping of the software to hardware. With occam this latter phase is less awkward than could be expected with other architectures. Whether or not these subsequent and important stages can be made more susceptible to mechanical development is an avenue of interest yet to be pursued, but certainly one deserving of attention in broadening the accessibility of parallel programming to the wider research community.

We have set out to investigate the feasibility of developing a multiprocessor-based system for use in auditory modelling adopting an approach which we believe is a promising line of inquiry. Central to all of the objectives identified is a recognition of the need to accelerate and improve the development process in a domain where high performance is essential and where flexibility in achieving this would be valuable. It has been shown that it is possible to develop concurrent systems and realize a design which integrates both digital signal processing and its support systems in one notation, and which can be targeted to hardware without necessitating redesign. Furthermore, we have demonstrated how some of the practical problems associated with deadlock and performance degradation can be tackled. Whilst there remains much to be done on better understanding the relationships between large aggregations of concurrent processes and the mapping of these on to hardware for performance purposes, it is clear that this avenue is one deserving of close investigation.

Acknowledgement

Acknowledgement is given to funding from the ESPRIT – Parallel Computing Action – Project 4147 towards the work presented.

References

Adamson, K. and Black, N. D. (1986) Physiological Modelling of the Human Cochlea. *IASTED International Symposium on Measurement, Signal Processing and Control*, Italy, pp. 53–66.

Inmos Ltd. (1988a) *Transputer Reference Manual*. Prentice-Hall International Series in Computer Science, Englewood Cliffs, NJ.

Inmos Ltd. (1988b) *Occam2 Reference Manual*. Prentice-Hall International Series in Computer Science, Englewood Cliffs, NJ.

McKeag, R.M. and Milligan, P. (1980) An Experiment in Parallel Program Design. *Software Practice and Experience*, **10**, p. 687.

Chapter 9

Modelling the Dispersion of Gaseous Pollutants in Air

J.R. Stewart

9.1 Introduction

This chapter describes an interesting problem where the most appropriate method of subdividing a problem for a transputer-based parallel solution is not obvious. The chapter presents models (and parallel implementations) of the dispersion of pollutants in air or water. Traditional models often make repetitive calculations over a two- or three-dimensional grid, and parallel algorithms are based on geometric division of the grid into subdomains. Similar code executes on a number of processors, and data from subdomain edges is exchanged at regular intervals. Such a solution is inappropriate for the example described here, and a 'pipeline' system was devised where different code works on a different part of the problem on each processor. Although the resulting program performs satisfactorily, for typical problem sizes the solution is not very efficient, and the inflexibility of the design is noted.

The introduction of legislation controlling the nature and extent of air pollution, the increasing awareness of environmental matters and problems associated with transnational air contamination have all contributed to increasing interest in the condition of the atmosphere. For over 30 years, meteorologists and other scientists have been developing methods to predict the transport and dispersion of gaseous pollutants in air. Turner's *Workbook of Atmospheric Dispersion Estimates* (Turner, 1967) and the *Handbook on Atmospheric Diffusion* (Hanna *et al.*, 1981) are examples of texts devoted to the practical aspects of calculating dispersion.

More recently, there has been a proliferation of computer programs produced to make the necessary calculations. Initially, these programs were written by atmospheric scientists for their own use, but latterly commercial companies have provided a wide range of programs for use by industrialists, planners and regulators.

The term 'air pollution modelling' actually includes a wide range of problems, each with its own particular solution. Different modelling systems may be required for short (up to about 100 metres), medium (up to about 20 km) and long range (up to hundreds of km) transport. There are different techniques suitable for modelling transport and dispersion from a factory chimney, or from a road system, or for the cyclical build-up and reduction of air pollution around a large city. In addition, it may be necessary to consider the half-life of radioactive pollutants, the reactivity of mixtures of chemicals or deposition in either dry or wet conditions.

There are various types of air pollution models, most of which use a combination of theoretical and empirical methods to predict transport and dispersion. The main ones are the Gaussian plume models, statistical models (Monte Carlo models), similarity models and gradient transport (K) models (see Hanna *et al.*, 1981, for detailed descriptions of these models). Each of these has been developed and modified by various researchers in an attempt to produce practical models which can cope satisfactorily with a range of conditions.

The project described here, and undertaken for the Queen's University of Belfast Environmental Science and Technology Research (Questor) Centre, uses a development of the Gaussian plume model to predict the dispersion of continuous or short-term releases into the atmospheric boundary layer. This layer varies in thickness from a few hundred metres at night to 2000 metres or more during day time. The Gaussian plume model may be used with reasonable confidence for distances of 10–20 km down wind from the source position.

This model is widely used, perhaps more because it has gained acceptance by various regulatory bodies, than because it is superior to other methods.

Currently available commercial programs use this model, among others, to provide simulation, monitoring and modelling facilities to users who either control or monitor processes which have routine or potential releases of gaseous pollutants. (For a summary of available programs and their features see Hanna and Drivas, 1987.)

Early dispersion programs were developed using the limited hardware and operating systems available at the time. They were generally rather slow, produced their output in the form of tables of numerical data, and were subject to restrictions on machine memory which often meant that the range over which the calculations were made or the grid size used for the analysis was limited.

With the improvement in available hardware, including the advent of high resolution colour graphics displays, various researchers have produced modelling programs using sophisticated enhancements to the original models and providing on-screen graphical results displays. At least one commercial company has produced programs which may be used in real time to predict the dispersion of emergency releases, and to be used as part of an emergency action plan (McPherson, 1989; Takacs, 1990). The Questor project aims to demonstrate the advantages of parallel processing in increasing the speed of programs and allowing reduced grid size in the analysis, and to produce a suitable design for a workstation incorporating the kind of sophisticated graphical user interface which may be built using modern windowing techniques.

9.2 The modelling system

This section begins with a description of the basic Gaussian plume model. This is followed by details of the puff model, which is derived from the plume model, and of the Questor modelling program.

9.2.1 The Gaussian plume model

The Gaussian plume model can be used to predict the concentration of a gaseous pollutant down wind from a fixed position source emitting at a steady rate. Figure 9.1 shows some of the basic parameters of significance in the calculation.

The emitted material is transported down wind, with velocity u, the local wind speed. At the same time the plume, which had initial dimensions determined, in part, by the exit diameter of the chimney, is increasing in both width and height. The growth is determined by local atmospheric turbulence. In turbulent conditions the material in the plume is dispersed more rapidly into the surrounding air, and the plume is spread more widely. Consequently, of course, if the same mass of material is distributed over a larger volume, the concentration of pollutant inside the plume will be lower.

The concentration at any point down wind from the source is calculated using Equation (9.1):

$$C(x,y,z) = \frac{Q}{2\pi u \sigma_y \sigma_z} * \exp\left(-\frac{y^2}{2\sigma_{y2}}\right) * \exp\left(-\frac{z^2}{2\sigma_{z2}}\right) \qquad \textbf{(9.1)}$$

Here Q is the release rate, u the wind speed in the direction of plume travel, y and z are horizontal and vertical distances from the centre line of the plume. Sigma y and sigma z characterize the dispersion rate, and may be estimated from actual measurements of lateral and vertical wind turbulence or, if such measurements are not available, from empirical tables which relate sigma

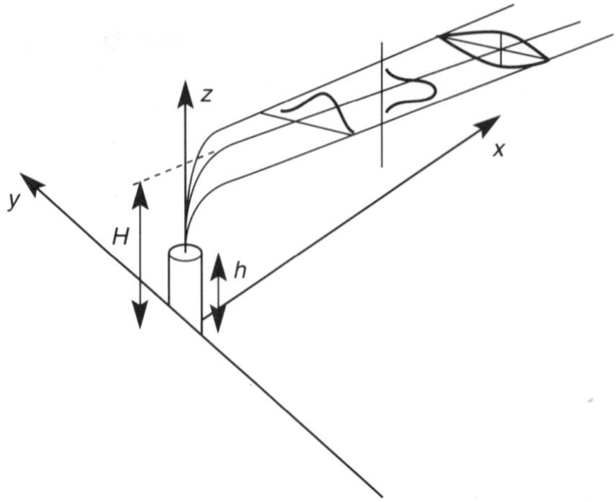

Figure 9.1 Some significant parameters used to describe emissions of polluting gases.

values to the travel distance from the source and the current atmospheric stability. The two exponential terms describe the Gaussian distribution of the polluting material about the horizontal (y) and vertical (z) axes.

9.2.2 The puff model

The plume model described above may be used to predict dispersion in a limited range of conditions. The model assumes a constant rate of release of the pollutant, a steady, uniform wind field, flat terrain, and will only work in conditions of transport, that is, with a minimum wind speed of about 1 m/second.

If material is emitted over a relatively short period, the dispersion characteristics may be modelled by considering the event as an instantaneous or **puff** release. In this case, all the emitted material is assumed to be distributed in an approximately spherical puff. The concentration calculation for puff releases is made using Equation (9.2):

$$C(x,y,z) = \frac{Q}{2\pi^{3/2}\sigma_x\sigma_y\sigma_z} * \exp\left(-\frac{x^2}{2\sigma_{x2}}\right) * \exp\left(-\frac{y^2}{2\sigma_{y2}}\right) * \exp\left(-\frac{z^2}{2\sigma_{z2}}\right) \quad \textbf{(9.2)}$$

The wind speed u is no longer used, and a new exponential term is introduced to describe the dispersion in the x direction. In this case, the concentrations calculated are for offsets in x, y and z from the current centre of the puff. Under transport conditions it is necessary to calculate the changing position of the puff centre from the current wind speed and direction.

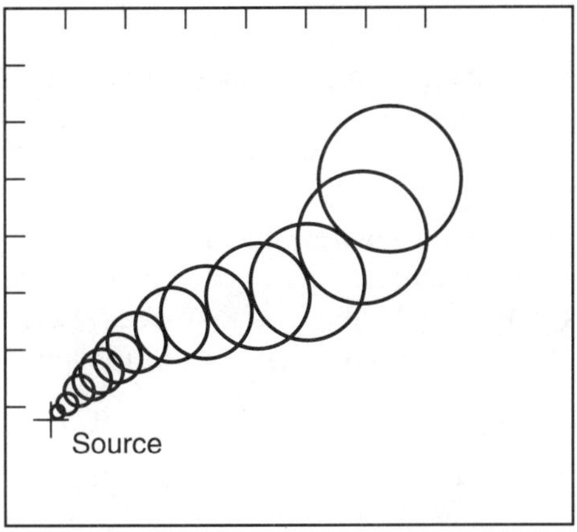

Figure 9.2 A continuous pollutant release represented by a chain of puffs.

It has been demonstrated (Ludwig *et al.*, 1977) that the results from the plume model can be represented with reasonable accuracy by considering the release as a number of discrete puffs, where the sizes of the puffs are predicted using the sigma values developed for the plume model. In reality, puffs and plumes disperse at different rates and their growth is described by different theories. Nevertheless, the technique has been adopted by many modellers. Releases over an extended period are then represented by a chain of consecutive puffs. The size of the calculation may be minimized by removing puffs which move out of the area of interest and by merging closely spaced puffs, when that is possible without significant loss of accuracy. The concentration of pollutant at any receptor position is calculated by summing the contributions from all puffs near enough to affect the result.

Use of the puff model confers several potential advantages. It allows convenient treatment of variable release rates, non-uniform and temporally variable wind fields and low wind or calm conditions. The effects of complex terrain features may be taken into account when estimating a wind field for use in the puff model, and this is also perceived as a useful advantage. Figure 9.2 illustrates the use of a chain of puffs to model a continuous release.

Comparative tests of various commercially available plume and puff models against experimental data have shown that, in at least half the cases, the models are substantially inaccurate, with concentration predictions differing from measured data by a factor of two or more. Also, more complex models do not show a significant improvement in reproducing reality in comparison with simpler models (Zanetti, 1986). In light of these inadequacies, much effort is currently being expended in attempts to refine more sophisticated models in a search for the necessary accuracy.

9.3 The Questor program

Several serial implementations of the puff model have been described in the literature (Zanetti, 1981, 1984; Ludwig *et al.*, 1977; Ludwig, 1984). For this project a serial version of the model was first constructed for testing, development and comparative purposes. The simplified flow chart in Figure 9.3 shows the major parts of the program and how they interact.

The time interval for the main puff generation and transport loop, shown here as ten minutes (a typical value), is determined by the frequency at which the wind field is updated. The puff generation/transport/dispersion loop is

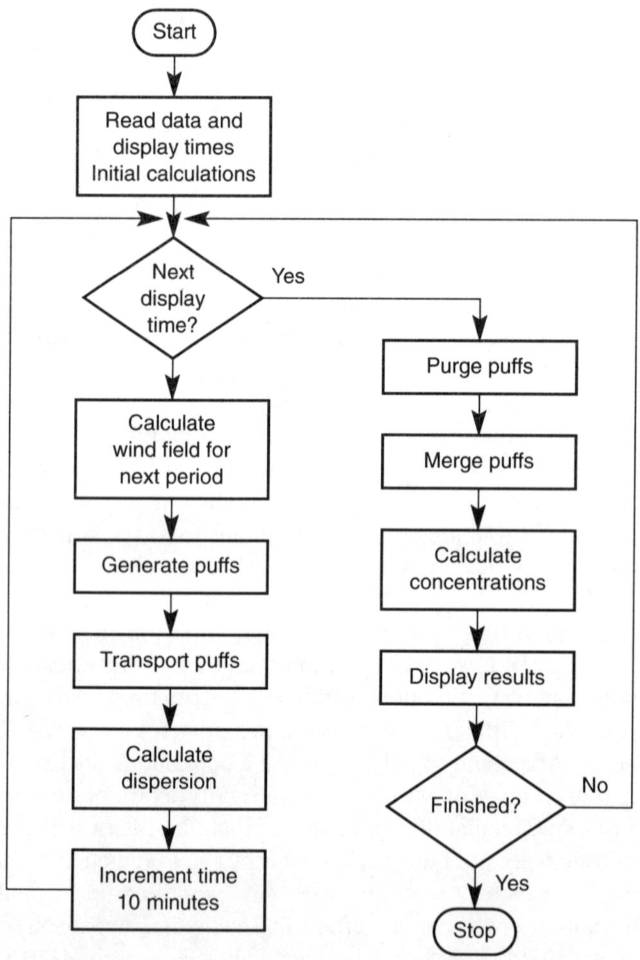

Figure 9.3 Flow chart for a sequential implementation of the puff modelling program.

repeated until the time advances to that requested for the display of the pollutant cloud. The next steps include removing puffs which have moved out of the area of interest, merging puffs where possible, calculation of the concentrations and display of the results. A record of the current puff chain is retained by the program, thus allowing users to run the model again to examine the results for later periods.

9.4 Parallel processing

9.4.1 The development system

The hardware available for the project was a Sun 3/260 workstation with a high resolution colour monitor and a Transtech NT1000 transputer support module fitted with four T800 transputers, each with 2 Mbytes of memory. The software system comprised v4.0 UNIX, the Sunview graphics library, Sun C compiler, Transtech's parallel operating system (the current system is known as GENESYS, but an earlier system was used for this project) and Transtech C for the transputer system, which includes extensions to implement various methods of message passing both between transputer nodes and between nodes and the Sun host processor.

9.4.2 An outline of the modelling program

The program will be used to model the transport and dispersion of releases of gaseous pollutants into the atmosphere. The user interface has been implemented using the Sunview windows system running on the Sun host processor. Users will enter appropriate details of the release and atmospheric conditions, and indicate the required format for display of results, the period of interest and the intervals at which concentration details should be calculated and displayed. The appropriate calculations will be made making the best use of available transputer processors.

9.4.3 A simple modelling program using the plume model

A common use of modelling programs is to calculate pollutant concentrations at each intersection of a grid covering the area of interest. For the plume model the concentration at any position in the grid may be calculated independently without the need to consider the concentration at other positions. If a grid of 100×100 cells is considered, then the 10,000 calculations may be split among any available processors. The C source code for the concentration calculation is shown below; it is derived from Equation (9.1):

```
void calculate_conc_grid ( )
    {
    double inv2sqsigmay , inv2sqsigmaz , temp1 , temp2 ,
        thissigmay , thissigmaz , zterm ;

    int  inx , iny , count , thisx , thisy , thisz ;

    thisz = 0 ;                   /* ground level concentration required */
    thisx = gridsize / 2 ;
    inx = 0 ;
    temp1 = sourcerate / ( pi * windspeed * 2 ) * 10e6 ;
    while ( thisx < xlimit ) {
        thisy = ylimit ;
        iny = 0 ;
        thissigmay = sigmay ( thisx ) ;
        thissigmaz = sigmaz ( thisx ) ;
        inv2sqsigmay = 1 / ( 2 * square ( thissigmay ) ) ;
        inv2sqsigmaz = 1 / ( 2 * square ( thissigmaz ) ) ;
        zterm = exp ( – square ( sourceheight – thisz ) * inv2sqsigmaz ) ;
        temp2 = temp1 / ( thissigmay * thissigmaz ) * zterm ;
        while ( thisy > 0 ) {
            conc_grid [inx][iny] = (int) ( temp2 *
                                exp ( – square ( thisy ) * inv2sqsigmay ) ) ;
            thisy –= gridsize ;
            iny ++ ;
        }
        count = 2 ;
        while ( thisy >= – ylimit ) {
            conc_grid [inx][iny] = conc_grid [inx][iny – count] ;
            count += 2 ;
            iny ++ ;
            thisy –= gridsize ;
        }
        thisx += gridsize ;
        inx ++ ;
    }
    }
```

This function was developed and optimized for a sequential version of the program. The sigma values are calculated once for each x offset and used for all the y offsets at that x value. In addition, the cloud is symmetrical about the centre line, and so the full calculation need only be made for the centre line position and one side of the centre line. In the absence of other considerations, a simple division of the task among a number of processors should take advantage of any such optimization. The task should therefore be divided along the x axis rather than the y axis. Four processors could calculate the 100 x by 100 y grid as 1 to 25 x by 100 y, 26 to 50 x by 100 y, etc. The load on each processor would not be quite equal. From the typical cloud shape shown in Figure 9.1, it is apparent that near the source the plume is much more narrow, and indeed, if the source is elevated it may not reach ground level immediately. In the function shown above, temp2 is the value of the centre line ground level concentration (in micrograms per cubic metre). If temp2 is zero, or as soon as the ground level concentration falls to zero, it is unnecessary to calculate any results for cells further away from the centre line.

In practice, it is unlikely to be worth trying to make the load on each transputer node exactly equal. Assuming there is a requirement to pass the results back to the host processor for assembly and display, and that the host can only accept one packet of results at a time, then it may be an advantage if the calculation processes finish at different times and do not have to queue to return their results.

A parallel implementation of this type is essentially straightforward. Identical calculation code is broadcast and initiated on each available transputer node. The host process must pass an appropriate set of parameters to each node available. Messages must be dispatched to a specific node address, and so the host must know how many nodes are available and their addresses. The task is divided evenly and a message passed to each node.

The route taken by any message is dependent upon the configuration of the system. The NT1000 board supports up to 16 nodes in four groups of four. It provides a flexible interconnection system, permitting a wide range of topologies which may be configured under software control during the boot sequence.

On receipt of the appropriate message each node completes the calculation required and passes the results back to the host. On the Transtech system, packet size is limited to 8192 bytes or 2048 integers for 4 bytes per integer processors. If the results array from any node exceeds this size it must be split and sent in several packets, each with an associated identification code. The various possible types of inter- and intra-node communication use a single parameter, a pointer to a network message descriptor:

```
struct nmsg {
        int nh_dl_event ;          /* for remote communications */
        int nh_node ;              /* destination node address */
        int nh_event ;             /* event code */
        int nh_type ;              /* message type, mask or filter */
        int nh_length ;            /* message length */
        int nh_flags ;             /* option flags */
        char * nh_msg ;            /* pointer to message */
};
```

Synchronization between sender and recipient nodes is enforced using the field nh_event, which must have the same value at both ends of the link. The sending process fills an nmsg structure and sends it to a specific address contained in nh_node. The receiving process waits to receive a message into a similar structure with the appropriate nh_event code. It is the programmer's responsibility to ensure that the nh_msg array on the recipient node is sufficiently large to accept the incoming message.

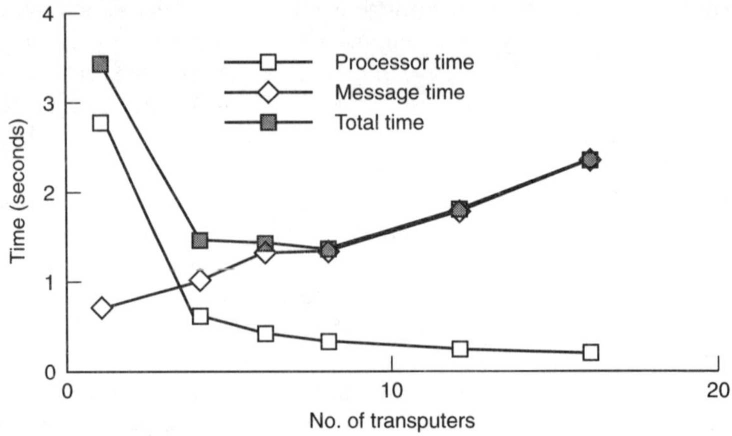

Figure 9.4 Processor and message passing times vs. number of transputers for the simple plume model.

A number of trials were undertaken using the method outlined above and varying numbers of transputers. The results, shown graphically in Figure 9.4, are for the calculation of a grid with 30,375 cells. A serial program running on the Sun host processor required 16 seconds for the same task. From Figure 9.4, it is apparent that, as the number of transputers increases, the processor time (per transputer) decreases. As might be expected, the product of number of transputers and processor time per transputer is almost constant. It is also noticeable that once the number of transputers exceeds eight, the total time is entirely dependent upon the rate at which messages can be transferred to and from the host processor. On the project system, message time normally averaged between 0.05 and 0.06 seconds/packet, irrespective of the size of the packet. (However, individual message times of the order of 1 second were occasionally recorded.)

For this somewhat unrealistic example, therefore, a near optimum performance can be achieved using four transputers. This test does serve to illustrate the point that this kind of message passing system is subject to severe limitation if large amounts of data are frequently to be moved around between nodes and the host. In certain circumstances, it is possible to improve the speed of message passing by establishing a connecting path between two nodes which may be held open for the transmission of multiple packets of data. This is only possible when no messages from other nodes interrupt and no link along the connecting path is required by another process. (See the Transtech manual, Chapter 4, for further details – Transtech, 1989.)

9.4.4 Modelling using the puff model

The puff model presents a more complex problem for both sequential and parallel programming. No calculation of the concentration at any position or time may be made until the entire chain of puffs has been generated and transported to the appropriate positions. At each time step (the ten minute intervals at which the wind field is updated) each puff in the chain is moved from its old position to a new one using the wind speed and direction at the current position of the puff centre. At each step a new puff size is calculated. Since puffs may potentially move in rapidly changing directions, or not at all if the wind speed is low, the entire puff chain must be scanned for each grid or receptor position to check whether any puff is sufficiently close to contribute to the concentration at that point.

A sequential version of the program, based on the puff model and the structure given in Figure 9.3, was written for comparative and testing purposes. Consideration was then given to the design of a parallel version. The calculation of a concentration grid, or concentration isopleths, on a number of transputer nodes will be constrained by the need to make available data for the whole puff chain at each node.

The introduction of parallelism may be made in various ways. In the simple example outlined above exactly identical code was executed on each available transputer node. However, this may not be the preferred solution for the more complex puff model. The diagram in Figure 9.5 shows how the various program tasks might be distributed, the aim being to complete as much

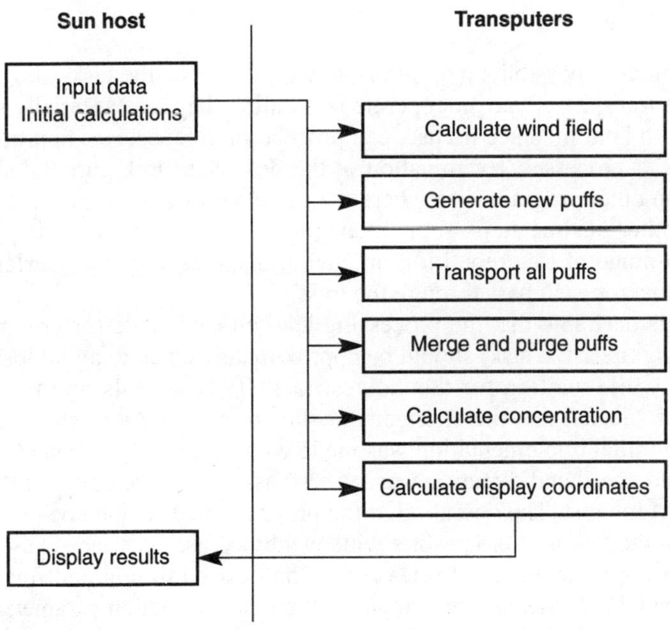

Figure 9.5 Distribution of processing tasks between the Sun host and transputer nodes.

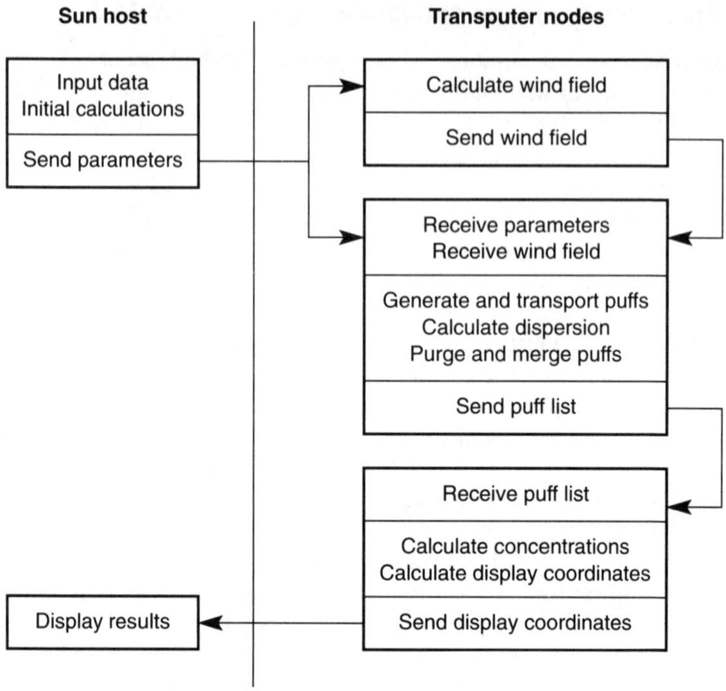

Figure 9.6 Trial apportioning of processing tasks between Sun host and transputer nodes.

processing as is possible on the transputers. Of course, the tasks shown as executing on transputers may not operate in parallel; they are essentially sequential from top to bottom, since no task can proceed until it receives appropriate data from earlier processes. Examination of the flow chart in Figure 9.3 shows that there is an outer program loop where the calculate-display sequence is repeated for a number of time steps after the beginning of the release. It is possible to take advantage of this repetition on each transputer node by overlapping the calculations for each pass through the loop.

It is desirable that the processing load on each node (or group of nodes performing the same task) should be approximately equal to avoid lost processing time while waiting for the slowest task. Timing trials on the sequential program indicated that the arrangement shown in Figure 9.6 would be reasonable. The initial implementation was made without the calculation of wind field data. Wind speed and direction was, at first, assumed to be constant throughout the area of interest. The operation of the programs is described below.

On the Sun host processor a windows-based user interface is used to enter details of a real or potential release and the desired timing and format of the results. Initial calculations are completed then the calculation parameters and list of display times is sent to node 2. The program then enters a loop to receive display coordinates from node 3, display the results and prompt the user to continue.

On node 2 the program waits to receive a set of parameters from the host. Once received (and determined to be calculation parameters rather than a quit message), control passes to a loop which, for each display time, completes the following operations:

> for each wind field period in the current display time generate, transport and calculate the size of the appropriate puffs, then,
>
> purge and merge puffs and send the puff list and parameters to node 3.

On node 3 the program waits to receive a puff list and parameters from node 2. Once received (and determined to be calculation parameters rather than a quit message), the concentration grid or isopleths are calculated. The coordinates for the screen display are then derived and sent to the Sun host processor.

9.4.5 An assessment of the puff model using transputers

The programs on the transputer nodes, once loaded and initialized, enter a continuous loop which waits for incoming data, calculates results and attempts to send them on to the next stage. Some time will be lost to prime the 'pipeline' at the beginning of a calculate-display cycle; the first display cannot be made until one complete (sequential) set of calculations has been completed. If the last operation in one cycle is to display the results on screen (and prompt the user to continue), and this is the slowest operation, then all the other processors in the system will have completed their work and be waiting to pass their results to the next node. After the first time through the loop there should always be a set of results available to the host for display on screen.

Timing trials carried out on the test system indicate that significant reductions in processing time can be achieved. The results for one set of tests are shown in Tables 9.1 and 9.2. These were obtained by running the programs over various calculate-display cycles ranging from 1–8 continuous cycles.

The timings in columns 2 and 3 in Table 9.1 were obtained by timing on the host processor between the dispatch of the initial parameters and the receipt of each set of display coordinates. All of the host processor times (for example, to display the screen image) have been excluded from these columns. Column 3 shows the cycle times when using two transputers and the program described in Section 9.4.4. For comparative purposes, all the transputer code was combined in one program running on one node. The results are shown in column 2. Column 4 shows the percentage reduction in transputer processing time when using two transputers instead of one. For the first two tests, one and two cycles, the time lost to prime the pipeline and on extra message passing negates all the benefits of splitting the code and operating in parallel. Thereafter the benefit is approximately 25%.

Table 9.1. Comparison of processing times (seconds) for the Sun host and the transputer system for a range of numbers of display cycles.

Number of cycles	*Transputers*		*Percentage reduction*	*Total times*		*Host alone*
	1 Tx	2 Tx		1 Tx	2 Tx	
1	0.40	0.40	0	2.04	1.86	3
2	0.94	0.90	4	3.24	2.84	9
3	1.56	1.20	23	6.12	6.44	4
4	2.22	1.76	21	7.86	7.20	20
5	2.88	2.08	28	9.30	9.00	26
6	3.78	2.84	25	10.96	10.34	32
7	4.72	3.56	25	13.32	12.26	38
8	6.18	4.74	23	14.86	14.20	47

Table 9.2 Comparison of transputer processor times (seconds) over eight calculation cycles using one and two transputers.

Cycle number	*Two transputers*		*One transputer*
	Tx 1	Tx 2	
1	0.210	0.170	0.356
2	0.300	0.193	0.468
3	0.378	0.193	0.538
4	0.399	0.204	0.587
5	0.406	0.214	0.580
6	0.616	0.211	0.779
7	0.668	0.217	0.818
8	1.158	0.205	1.316
Total	4.135	1.607	5.442

The last three columns in Table 9.1 show the total processing time, including screen drawing, using the host alone and one and two transputers. There is obviously a considerable advantage in using even one transputer operating in parallel with the host. The difference between using one and two transputers is almost insignificant.

The timing details in Table 9.2 were obtained by timing on the transputer nodes, from the receipt of parameters to the dispatch of results, for each cycle of an 8-cycle run. Considering the totals row, if the processing times for the two transputer test overlap, the decrease in processing time is from 5.442 to 4.135 seconds, or about 25%. Examination of the processing times in columns 1 and 2 shows that the first transputer node always dominates the total processing time. For the eighth cycle the ratio of first to second times is about 5:1. An improved design would spread the load more evenly.

9.5 Further developments

As mentioned above, the test system assumes a fixed wind field. The calculation of a wind field for the area of interest, either by interpolation of data collected from surrounding weather stations, or by calculations based on local meteorological measurements and details of the surrounding terrain, is a complex and time-consuming operation. Such calculations would add considerably to the processing time for a sequential system. It is anticipated that by dedicating two or more additional transputers to this task the overall times shown above for the transputer system would show only marginal increases.

The introduction of a buffering system for interprocess communication is currently under investigation. For the system described above a receiving buffer on the host processor of 8 Kbytes per cycle would generally be sufficient. On the transputer nodes even less storage would be required. Without a buffering system none of the processing nodes may complete the dispatch of a message until the receiving node is able to accept it. If the host is not ready to accept display coordinate data then each of the other nodes will eventually be blocked waiting to send. If the host could accumulate sets of display data in a buffer then the transputer nodes could complete the remainder of any calculation while the user is examining displayed results. The accumulation of results on the host processor would make it a straightforward task to allow the user to view them in any chosen order.

There still remains the possibility of dividing up any of the processes identified in Figure 9.6 over more than one transputer. As shown in Table 9.2, one of the transputer nodes dominates the total time for processing on the transputers. It should be possible to reduce the load on this first transputer either by splitting the code among more transputers or by running similar code on other transputers and splitting the data processing among them.

Consideration of how to achieve further increases in speed raises the question of uprating the performance by adding further transputers to the system. Now it becomes apparent that using a pipeline design is a distinct disadvantage. The simple system outlined in Section 9.4.3 for the plume model could run unaltered on any number of available transputers (although not with equal efficiency). The program described here lacks that flexibility, although from a practical point of view the objective of reduced processing time has been achieved.

9.6 Related areas where parallel processing might be used

The calculation of a wind field grid for the area of interest has already been noted as an important part of dispersion modelling. The methods used are mathematically complex, often involving the solution of differential equations. The use of finite difference and finite element methods for deriving wind fields has been described by various researchers (Bhumralker *et al.*, 1979; Sherman, 1978), and there is scope for improving processing times in these computationally intense methods.

There is also much current interest in modelling flow regimes and pollutant distribution in aqueous systems such as rivers, estuaries and seas, and in land-based situations for the dispersion of pollutants, such as the leachate from land fill sites, through the ground strata and into water tables. Such systems are often modelled using finite element analysis. The systems to be modelled are generally large and complex, and it is expected that parallel processing methods will make a very useful contribution.

Acknowledgement

The project described in this chapter was carried out for the Queen's University Environmental Science and Technology Research (Questor) Centre at the Queen's University of Belfast.

References

Bhumralker, C.M., Ludwig, F.L. and Mancuso, R.L. (1979) *Estimation of wind characteristics at potential wind energy conversion sites.* Final Report BNL sub-contract B-29105-A-E. (Available from NTIS, Springfield, Virginia.)

Hanna, S.R., Briggs, G.A. and Hosker, R.P. (1982) *Handbook on Atmospheric Diffusion.* Technical Information Centre, U.S. Department of Energy.

Hanna, S.R. and Drivas, P.J. (1987) *Guidelines for use of Vapor Cloud Dispersion Models.* American Institute of Chemical Engineers.

Ludwig, F.L. (1984) Transport and diffusion calculations during time- and space-varying meteorological conditions using a microcomputer. *J. Air Pollution Control Assoc.* **31** (3), pp. 227–232.

Ludwig, F.L., Gasiorek, L. S. and Ruff, R.E. (1977) Simplification of a gaussian puff model for real-time minicomputer use. *Atmospheric Environment* **11**, pp. 431–436.

Ludwig, F.L. and Byrd, G. (1980) An efficient method for deriving mass-consistent flow fields from wind observations in rough terrain. *Atmospheric Environment* **14**, pp. 585–587.

McPherson, S.W. (1989) Columbia Nitrogen's Spill Alert Computer System. *Plant/Operations Progress* **8** (4), pp.206–208.

Peterson, W.B. and Lavdas, L.G. (1986) *INPUFF 2.0 – A multiple source gaussian plume dispersion algorithm.* Report EPA – 600/8-86-011 U.S. Environmental Protection Agency.

Sherman, C.A. (1978) A mass-consistent model for wind fields over complex terrain. *J. Applied Meteorology* **17**, pp. 312–319.

Takacs, L. (1990) Episodic Modeling for Emergencies. *Pollution Eng.* January.

Transtech (1989) *Niche Parallel Solutions Manual.* Transtech Devices Ltd., High Wycombe, UK.

Turner, D.B. (1967) *Workbook of Atmospheric Dispersion Estimates.* U.S. Public Health Service Publication No 999-AP-26.

Zanetti, P. (1981) An improved puff algorithm for plume dispersion simulation. *J. Applied Meteorology* **20**, pp. 1203–1211.

Zanetti, P. (1986) A new mixed segment-puff approach for dispersion modeling. *Atmospheric Environment* **20** (6), pp. 1121–1130.

Chapter 10
Transputer Database Systems

M.E.C. Hull and J. Shao

10.1 Introduction

This chapter gives an overview of the issues related to parallelism in database systems. It then proceeds to illustrate how transputers can be used as building blocks in developing parallel database systems. The current state-of-the-art in parallel relational database systems is first surveyed. Attention then turns to transputer database machines. Parallel implementation of the key relational database operators on a transputer-based machine is considered in some detail.

Database systems have become standard tools for hiding the details of secondary storage management from computer naïve end users, and hence facilitating data access. They offer a much higher level of data independence, a greater flexibility and a better facility for sharing data than the traditional file systems. The high level query languages supported by these systems make it possible for applications to be coded in days rather than in weeks or months when using file systems. Database applications can now be found in various areas, from airline seat reservation systems to medical information systems.

However, many applications still use simple files. To a great extent, this is because of the performance of current database systems; they are not offering the efficiency that users require for their applications. Despite the increase in processing power, the performance of database systems continues to be an important and critical issue. This is because of the following:

- The volume of data is ever increasing. Already there are applications where more than 100 Gigabytes of data are required to be stored in a single database and to be kept on-line all the time. The growing applications in non-traditional areas, such as VLSI, computer-aided design (CAD) and computer-assisted software engineering (CASE), are expected to involve terabytes of data.
- User queries are becoming more complicated. Supported by very high-level query languages, end users can now easily code a fairly complex query. In some cases, they may not even be aware of the complexity of their requests.
- The number of on-line users is increasing. The advance of computer communication and network technology has made it possible for a database to be shared not only locally, but also nationally and even internationally. An airline seat reservation system, for example, may need to respond to thousands of user requests simultaneously.

To achieve further improvement in database system performance, parallel processing techniques become essential and attractive, particularly because parallel processors have now become commercially available and affordable. In recent years, there have been several proposals, prototypes and commercial systems that make use of parallel architectures. In this chapter, we shall give an overview of the issues related to parallelism in database systems. We shall also illustrate how transputers can be used as building blocks in developing parallel database systems.

10.2 Relational database systems

If we accept the fact that a relational data model is the state of the art having received much recent attention from both researchers and database system developers, we can then consider the parallelization of relational database systems. To make this chapter self-contained, a brief introduction to the basics of relational database systems is given.

A relational database consists of a set of relations. Intuitively, each relation can be viewed as a table of values. Names are often given to the columns of the table. These names are called **attributes** of the relation. Each row of the table is a **tuple** of the relation. Table 10.1 gives an example of a relation called **Parent**, viewed in tabular form.

As the name suggests, the *Parent* relation stores information about the parent relationship. *Pname*, *Cname* and *Psex* are its attributes. The tuples of this relation, say the first row <*John, Fred, Male*>can be interpreted as 'John is a male and has a child called Fred', or simply 'John is Fred's father'.

Table 10.1 *Parent* relation.

Pname	*Cname*	*Psex*
John	Fred	Male
Fred	Mary	Male
Mary	Tom	Female

Often, we need not be concerned with the actual values stored in a relation in our discussion. In such cases, a simple schema representation of a relation will be used. For example, the *Parent* relation given in Table 10.1 can be denoted by the schema:

parent (Pname,Cname,Psex)

that is, we denote a relation only by its name and attributes.

Users access a relational database through relational query languages. While the formation of queries in these languages may vary, depending upon the nature of the problem and the query language used, their execution is similar. First, a **query tree** is generated for a query. The nodes of the tree represent the operations required to evaluate the query, and the edges represent the order in which the operations are executed. Thus, a query tree serves as an execution plan for the query. According to this plan, the execution of the query is then carried out. The execution will start at the leaves of the tree and the result obtained will be sent to its parent node. This continues until the operation at the root of the tree is executed.

The nature of a query tree (its size and shape) depends upon the complexity of a query. The operations that are often used in a query tree include standard relational algebra operations, such as *select* and *join*, and low-level supporting operations, such as *sort* and *index*. Later in the chapter we will examine in detail how a query tree can be mapped onto a network of transputers for parallel processing.

10.3 Parallelism in relational database systems

In this section, we consider how parallelism may be exploited in a relational database system. There are four main forms of parallelism which can be addressed with respect to the development of a parallel database system.

10.3.1 Transaction parallelism

One of the main purposes of a database system is to facilitate data sharing. A number of users can access (query and update) a single database simultaneously, yet leaving the impression that they are the sole user of the system. In database terminology, each individual access made by a user is called a **transaction**, and it is clearly desirable to have transaction parallelism. That is, independent transactions submitted by users should be executed in parallel, possibly on multiprocessors.

Transaction parallelism is similar to the traditional multi-tasking in operating systems. They are particularly beneficial in terms of enhancing system throughput. Care must be taken, however, because these 'independent transactions' are not truly independent; they all access a single, shared database. Thus, mutual exclusion and must be ensured and deadlock must be avoided. These are normally accommodated in a database system by a component called **concurrency control**.

10.3.2 Function parallelism

In this form of parallelism the system function is the main attribute. A database management system can be divided into several functional components, each of which can be placed on a separate processor so that different functions are performed in parallel across multiple processors. For example, user interface, query evaluator and disk access controller can be assigned to different processors.

Function parallelism is particularly useful when it is used in conjunction with transaction parallelism. Although users may access a database simultaneously, they may require different functions from the system. Thus, the provision of function parallelism can be useful in this way. However, the main drawback with this form of parallelism is that the speed-up achievable is limited because the number of different functions that can be identified and assigned to multiprocessors is limited.

10.3.3 Operation parallelism

Operation parallelism is concerned with the possible exploitation of parallelism in evaluating a single query. Clearly, this is equivalent to the question of how to execute the corresponding query tree in parallel. Two strategies are possible. We illustrate these using the query tree given in Figure 10.1, where Op_1, Op_2 and Op_3 are three operators, and R_1, R_2, R_3 and R_4 are four source relations.

The first strategy assigns each operator in the tree to a separate processor for parallel processing. Thus, the example given in Figure 10.1 will use three processors in processing the query. Some communication (data exchange) will also be needed. However, it should be noted that although three processors are used, the

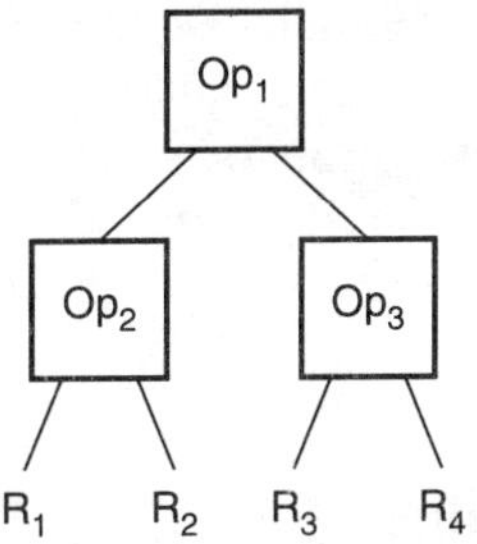

Figure 10.1 A query tree involving three operations.

degree of parallelism may not be three. This is because the one with Op_1 assigned to it may not start execution until its child operations are all completed.

The above observation has led to the use of second strategy, namely the pipelining technique. Using this technique, the partial results (tuples) output by one operation can immediately be consumed by another operation, while the two operations are being executed concurrently. In Figure 10.1, for example, as operation Op_2 produces tuples, they can be sent immediately to the executing operation Op_1 which will absorb them as they arrive. As such, the degree of parallelism may be increased.

Generally speaking, the degree of parallelism obtainable by this form is limited by the number of operators used in the query tree, which reflects the complexity of a user query. Furthermore, the actual degree of parallelism attainable is usually much less than this upper bound due to the dependencies among the operators. Thus, in most cases exploiting operation parallelism alone is not sufficient to achieve great speed-ups. However, it is useful when it is used in conjunction with data parallelism, which is discussed below.

10.3.4 Data parallelism

Data parallelism is the key to supporting a high degree of parallelism in a database system. In this form of parallelism, each processor evaluates a copy of the same query (or a copy of the same operation if operation parallelism is used), but with less data. This is achieved by dividing a relation into a number of partitions and assigning each partition to an individual processor.

Obviously, data partitioning strategies play an essential role in using this form of parallelism. A good partitioning strategy should maximize local processing and minimize interprocessor data exchange. That is, a relation should be so partitioned that only the tuples at the local site will have the possibility to produce results. Unfortunately, this is generally difficult to achieve.

In the above sections, we addressed four main forms of parallelism in database systems. It is interesting to note that they actually represent parallelism at different levels: transaction parallelism at user level, function parallelism at system level, operation parallelism at query level and data parallelism at database level. Ideally, all these forms of parallelism should be exploited in a parallel database system. In the following sections, we shall examine closely how these forms of parallelism may be used practically.

10.4 Database machines

The development of parallel database systems is largely represented by the work in trying to migrate software functions of database systems into hardware. Since the mid 1970s, a great deal of effort has been directed towards the design of special-purpose architectures for efficient handling of large database systems. The term **database machine** is often used to refer to this class of database system, consisting of special hardware and software components. It is hoped that by having parallelism exploited by the hardware, these database machines will perform more efficiently than their software equivalent. In this section, we shall give a brief introduction to two representative database machines, thereby providing some insight into how parallelism may be used in database machines in general.

10.4.1 DIRECT

DIRECT (De Witt, 1979) was the first database machine which completely implemented the functions of a relational database system. DIRECT is a collection of query processors and memory modules interconnected through a crossbar switch network. The architecture of DIRECT is shown in Figure 10.2.

The network provides connections between any pair of query processor and memory module. Query processors are PDP11/03 with 28 K word memory

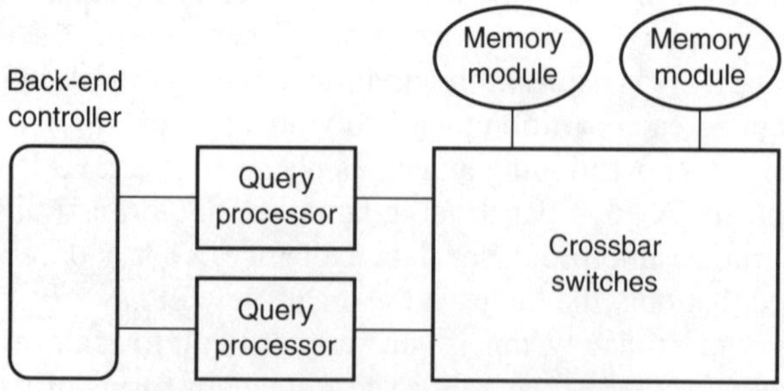

Figure 10.2 DIRECT configuration.

and memory modules are shared disk caches. The back-end controller (PDP11/40) is responsible for determining the number of query processors and memory modules that should be allocated to a query. This is done based on the complexity of the query, size of the database and the system contention.

In this design, the back-end is mainly used for query processing. Thus, mainly the operation parallelism and data parallelism are used. However, the front-end (PDP/1145 running under UNIX) supports transaction parallelism. The design offers good load balancing and allows the processors to be shared among the storage modules. Furthermore, expansion is easy and modular. However, it was soon discovered that a lot of data needed to be transferred from one module to another, causing bottlenecks at the crossbar, and hence degrading the performance.

10.4.2 GRACE

GRACE (Kitsuregawa *et al.*, 1984) represents a different strategy in designing a database machine. While parallel versions of **Brute Force** algorithms are used in DIRECT to evaluate relational algebra operations, GRACE is aimed at achieving high performance by adapting processing algorithms based on hash and sort operations. The general philosophy behind the query processing in GRACE is based on (1) clustering the database via hashing, (2) uniformly distributing clusters among several temporary storage units, and (3) processing the distributed clusters in parallel. The overall configuration of GRACE is shown in Figure 10.3.

The system consists of two rings (processing ring and staging ring) and four modules (processing module (PM), memory module (MM), disk module

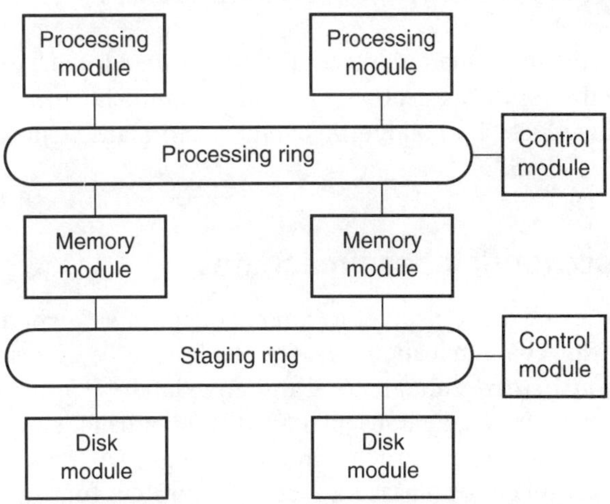

Figure 10.3 GRACE configuration.

(DM) and control module (CM)). The processing module performs relational algebra operations on the data stored in the memory modules in a pipeline fashion. The disk modules are small processors capable of filtering the data according to the user query. Finally, the control module provides control in the system. All the modules are special-purpose silicon chips.

Similar to DIRECT, the back-end of GRACE is mainly used for query processing, and thus mainly the operation parallelism and data parallelism are exploited. However, by separating disk modules from processing modules, GRACE does offer a certain degree of function parallelism. Finally, GRACE is also a multi-user system. The most attractive feature of the GRACE design is its algorithm for join, a relational algebra operation. By proper distributing (through hashing) and pipelining, the system is able to process the join operation with great efficiency.

From the above brief discussion on database machines, we make the following observations:

(1) Modularity in architecture is essential to a database machine. Both DIRECT and GRACE have shown to be modular in their architectures. This is particularly important if a database machine is to be **scalable**, that is, when the workload increases, the performance of the database machine will not be degraded because additional processing elements may be **plugged** into the system easily, thereby providing further processing power.

(2) Proper data distribution is essential to a database machine. The problem revealed by DIRECT and the success of the GRACE join algorithm are clear support for this need. Relations should be partitioned so that their parallel processing could be carried out largely independently, that is, without excessive interprocessor communications.

Clearly, the transputer architecture is well suited for achieving these two objectives. In the following section, we shall examine how transputers can be used as building blocks for configuring a high performance database machine.

10.5 Transputer database machines

Useful properties of the transputer as a component in a database machine are speed, granularity, communication capacity and flexibility. The transputer architecture is *sophisticated* enough to allow any database functionality to be implemented even on a single transputer. On the other hand, it is *flexible* enough to allow a network to be constructed in any topology. A transputer communicates with its neighbours by passing messages through four fast bidirectional serial links. While it is not possible to discuss at full length in this chapter the

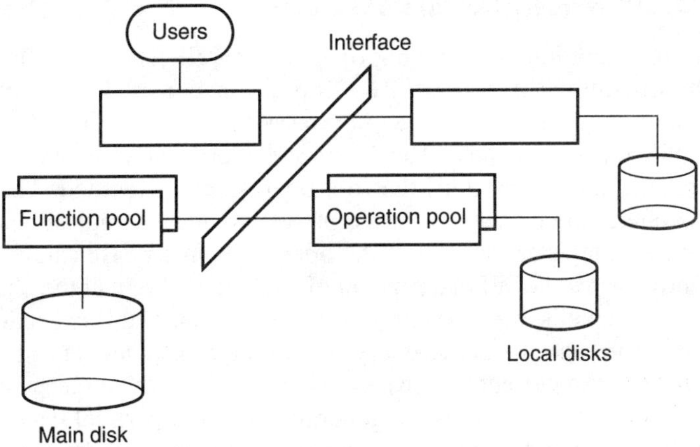

Figure 10.4 A simple transputer database machine architecture.

various ways in which transputers may be used to configure a database machine, we focus on a simple and abstract architecture shown in Figure 10.4.

The **function pool** consists of a set of transputers, each performing a particular function. For example, we may have one transputer responsible for user query evaluation (that is, translate a query into a query tree and then map it onto the operation pool), and another responsible for backup and recovery (in case there is a system crash). If it is a multi-user environment, we may also have one transputer providing the required concurrency control. All these functions are performed in parallel, and hence function parallelism is directly supported in this architecture.

The **operation pool** consists of a set of transputers, each associated with its own local disk. As the name suggests, these transputers are responsible for performing relational and other operations in parallel. The operations are received as **commands** from the function pool, through the interface which is a specially designed hardware component. Using a special hardware component is a common practice in constructing database machines, and often enhances their performance. Operation parallelism is achieved by mapping a query tree onto the transputers in this operation pool, and data parallelism may also be exploited by partitioning data among these transputers.

In what follows, we focus on the function of the operation pool because it is the main source for achieving a high degree of parallelism. That is, we shall address the issues related to how a query tree should be mapped onto the transputers in the operation pool; how the relational and other operations should be performed in parallel across these transputers and how data should be partitioned. For the purpose of this chapter, we shall keep our discussion on these issues as general and simple as possible.

10.5.1 Mapping query tree to transputers

The key consideration in achieving operation parallelism is to allocate optimally the operations in a query tree to the transputers. For simplicity, we assume that a query tree is executed in **phases**. In the first phase, the leaf operations of the tree are executed. Then their parent operations are executed in the second phase, and so on, until the root operation is executed. The order in which the phases are executed is strict: phase $i-1$ must complete its execution before phase i may start. Note that the operations in a phase can be executed concurrently as they are all independent of each other. We call the phase which is currently executable the **current phase**. Clearly, under this execution model, it is pointless to map the entire query tree to the transputers at once. Instead, the optimal mapping of current phases, one at a time, to the transputers should be considered. This can easily be generalized to the parallel execution of n query trees, corresponding to n user queries. In that case, all the current phases of each tree form a global current phase and this **global current phase** is then mapped to the transputers in a similar way.

Given a current phase consisting of n operations $<O_1, O_2, ... , O_n>$ and m transputers $<T_1, T_2, ..., T_m>$, one way to allocate $<O_1, O_2, ... O_n>$ to $<T_1, T_2, ... T_m>$ is to use a *one-to-one* mapping, that is, one operation is mapped to one transputer. This is simple, but may not be optimal. For example, when $m > n$ the system resources will be under-used. Even when $m = n$ each operation in the phase may not require the same processing power. As a result, some transputers will finish their execution earlier than others. This may seriously limit the performance as the execution of the next phase will not start until the execution of all the operations in the current phase is completed.

To improve the mapping, a very simple model based on the **mean operation time** for a phase may be used (Leung *et al.*, 1990). Let the total time required for computing $<O_1, O_2, ..., O_n>$ on a single processor be:

$$TT = \sum_{i=1}^{n} t_i$$

where t_i is the time required to complete O_i. With m transputers available, it is intuitively fair that each transputer should contribute the following amount of processing time to the execution of the phase:

$$MT = \frac{TT}{m}$$

MT is thus the **mean operation time** for the phase. Using this mean operation time as a benchmark, the number of transputers which should be allocated to the execution of O_i is given by:

$$k_i = \frac{t_i}{MT}$$

Thus, a simple operation (that is, $t_i \approx MT$) should be assigned to a single transputer, whereas a more complex operation (that is, $t_i \gg MT$) will use several transputers. In that case, data distribution must be considered. Also, there may be cases where we need to map more than one operation onto one transputer. This is often due to the fact that the operations in a given phase are **unbalanced.** Some operations are much simpler than others and thus take much less time to compute. For example, it is possible to have $\mathrm{k}_i + \mathrm{k}_j \approx 1$ for two operations O_i and O_j, $i \neq j$, in a given phase. It is clearly sensible and useful to map both O_i and O_j onto the same transputer for execution under this condition.

Note that in the above model, we have effectively assumed that the time required for computing each operation (that is, t_i for O_i) is known. Strictly speaking, precise t_i for O_i cannot be known until O_i is executed. So in practice, t_i is obtained largely through estimation or approximation. This is usually a fairly complicated task, and is dependent upon such factors as implementation algorithms, parameters and statistics of the involved relations. There are various methods available, but we shall not discuss the matter further in this chapter.

10.5.2 Parallel join

Join is one of the most fundamental operations used in relational database systems. Let R and S be two relations, each having k_1 and k_2 attributes, respectively. The *join* of R and S over attributes i (of R) and j (of S), written as $R \bowtie 1_{i\theta j} S$, is a set of all possible tuples t that satisfy the following:

- the first k_1 values of t form a tuple t_1 in R;
- the last k_2 values of t form a tuple t_2 in S;
- θ is a comparison operator;
- the ith value of t_1 and the jth value of t_2 satisfy θ.

The following is an example of join operation.

A	B	C
1	2	3
4	5	6
7	8	9

$\bowtie$ B < D

D	E
3	1
6	2

=

A	B	C	D	E
1	2	3	3	1
1	3	3	6	2
4	5	6	6	2

The join operation is the most expensive operation to perform in a relational database system. There has been a great deal of effort in trying to devise efficient algorithms for computing the join operation. In this section, we consider a straightforward parallelization of a simple, brute-force join algorithm called **nested loop**, thereby showing the benefit of parallel processing.

Let R_1 and R_2 be two relations to be joined. The nested loop algorithm for computing $R = R_1 \bowtie_{a1 \theta a2} R_2$ is given below. The program is coded in occam 2. For simplicity, we have assumed that both relations are of integer type and some operations are expressed in an abstract notation. For example, a comparison operation is represented by <θ>:

```
PROC Nested.Loop ([][] INT R1, R2, R, INT a1,
                                        a2, <θ> )
  INT i, j, k:
  SEQ
    k := 0
    SEQ i = 0 FOR SIZE R1
      SEQ j = 0 FOR SIZE R2
        IF
          R1[i][a1] <θ> R2[j][a2]
            SEQ
              k := k + 1
                  R[k] := R1[i] <concatenate>
                                             R2[j]
          TRUE
              SKIP
:
```

Notice that when this algorithm is used to compute the join operation, it requires $s_1 \times s_2$ comparisons and s concatenations, where s_1, s_2 and s are the sizes of R_1, R_2 and R, respectively. This can be very time-consuming if relations are large.

To parallelize this simple nested loop join algorithm, the two involved relations, R_1 and R_2, are partitioned across k transputers. The smaller of the two joining relations, say R_1, is partitioned first through a **split table** which contains an entry for each transputer. A hash function is applied to the join attribute of each tuple. The value of this function maps the tuple to an entry in the split table, and thus determines to which transputer the tuple should be distributed. As the tuples arrive at a transputer they are stored in the local disk attached to the transputer. The other joining relation R_2 is then distributed in the same manner. Note that when a relation is distributed in this way, its subrelations are stored in each local disk in parallel. Following that, a local nested loop join is performed on two subrelations. The program which runs on each transputer is given below:

```
PROC Local.Join (CHAN from.function.pool,
                              to.function.pool)
  [rows][columns] INT R1, R2, R:
```

```
    INT a1, a2:
    SEQ
      from.function.pool ? R1, R2
      from.function.pool ? a1, a2, <θ>
      Nested.Loop (R, R1, R2, a1, a2, <θ>)
      to.function.pool ! R
  :
```

The set of `Local.Join` processes, each running on a separate transputer, is then executed in parallel. The following occam code expresses the parallel version of the nested loop algorithm:

```
[number.of.transputers]CHAN from.function.pool,
                                 to.function.pool:
PAR i = 0 FOR number.of.transputers
  Local.Join (from.function.pool[i],
                             to.function.pool[i])
```

where `from.function.pool` and `to.function.pool` are sets of channels linking the function pool to the transputers in the operation pool. Each transputer T_i receives subrelations and other parameters from the function pool through channel `from.function.pool[i]` and returns the result to the function pool via channel `to.function.pool[i]`.

Notice that since the same hash function is used to distribute both relations, only the tuples within the subrelations at a particular transputer have the possibility of joining. Thus, independent local joins performed in parallel will fully compute the join. However, the complexity is reduced. It is clear that under this parallel evaluation, each transputer T_i is required to perform $s_{1i} \times s_{2i}$ comparisons and s_i concatenations, where s_{i1}, s_{i2} and s_i are the sizes of the subrelations on T_i. Thus, if the data distribution is even and the hash function is fair, it is possible to have $s_{i1} \approx s_{j1}$, $s_{i2} \approx s_{j2}$ and $s_i \approx s_j$, for all $1 < i,j < m$. This will result in a speed-up factor of m, the number of transputers used.

10.5.3 Sorting operation

Although it is not directly available to users, sorting is used extensively in a relational database system as a support operation for implementing other operations. For example, instead of using the brute force nested loop algorithm, join may be implemented more efficiently using a **sort-merge** algorithm, where relations are sorted first and then merged into the final result. Thus, an efficient sorting algorithm is essential to the performance of a relational database system.

For simplicity again, we consider a simple sorting algorithm called **bubble sort**. The algorithm is given below, where a is the attribute on which relation R is to be sorted:

```
PROC Bubble.Sort ([][] INT R, INT a)
  [columns] INT t:
  INT i, j:
  SEQ i = 0 FOR SIZE R
    SEQ j = i + 1 FOR (SIZE R) - i
      IF
        R[i][a] > R[j][a]
                SEQ
                  t := R[i]
                  R[i] := R[j]
                  R[j] := t
                  TRUE
                    SKIP
:
```

The simplest method to sort a relation in parallel is the sort-merge algorithm.

Here a set of transputers are arranged in a tree structure, as shown in Figure 10.5, and the algorithm works in two steps. In the first step, data is sorted locally at each leaf node of the tree and the sorted data is forwarded to the parent node. In the second step, each non-leaf node of the tree receives two or more streams of sorted data from its child nodes. These streams of data are compared and merged, with proper ordering, into one stream. The new stream is then forwarded again to its parent node. The process is repeated until the root of the tree has merged all the data streams from its children.

Each leaf node basically performs the bubble sort algorithm, plus some communication. The program is given below, where we have assumed that relation R is partitioned, before the execution starts, into R_1, R_2, ... , R_m, and m is the number of leaf nodes. Variable R_i is used in the program to hold subrelation R_i:

```
PROC Leaf (CHAN to.parent, [][] INT Ri, INT a)
  INT i:
  SEQ
    Bubble.Sort(Ri,a)
    SEQ i = 0 FOR SIZE Ri
      to.parent ! Ri[i]
    to.parent ! Max
:
```

Note that we have used **fine-grained parallelism** in the above program, that is, each leaf node sends the sorted subrelation to its parent node tuple by tuple. As such, the parent node is able to consume a tuple as soon as it is produced from its child nodes. A fine-grained pipelining often increases the degree of parallelism. It is possible, and sometimes necessary, to send a group of tuples together if communication overhead is becoming too severe. *Max* is a special constant which must be greater than any value in relation R. This special constant signifies the end of a sorted stream.

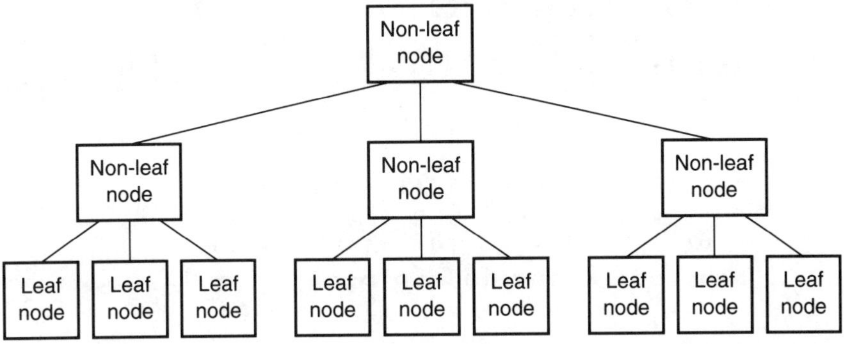

Figure 10.5 The tree-structured sort-merge algorithm.

In describing the behaviour of the non-leaf nodes of the tree, we assume that each one has *f* children. The number of children that a node can have is often called the **fanout** of the tree. The program for non-leaf nodes is given below:

```
PROC Non.Leaf (CHAN to.parent,
                              [] CHAN from.children)
  [f][columns] INT streams:
  INT i:
  SEQ
  PAR i = 1 FOR f
    from.children[i] ? streams[i]
  WHILE TRUE
    SEQ
      i := min (streams)
      to.parent ! streams[i]
      from.children[i] ? streams[i]
:
```

So the parent node 'sees' the smallest values of *f* sorted streams of data from the children nodes. These values are compared and the smallest one is forwarded to the parent node. The stream containing the tuple that has just been output is replenished and the process of comparison is repeated. This continues until the special constant *Max* is received from all the child nodes.

Now we can put all the processes together. Assume that the height of the tree is *h* and fanout *f*. Then:

$$n = \sum_{i=1}^{h} f^i$$

is the number of communication channels needed to link the nodes of the tree together. Also, assume that the communication channels are numbered top-down and left to right (as shown in Figure 10.5). Then the program for the sort-merge algorithm can be given below:

```
PROC Sort.Merge ([][] INT R, INT a)
  [n] CHAN pipe:
  INT i:
  SEQ
    partition (R, R1, R2 ..., Rfh)
  PAR
    PAR i = (fh - 1)/(f - 1) FOR fh
      Leaf (pipe[i], R_, a)
    PAR i = 0 FOR (fh - 1)/(f - 1)
  Non.Leaf (pipe[i], [pipe[i × f + 1],...,
                            pipe[i × f + f ]])
:
```

Here, the function `partition` partitions relation R into f^h (the number of leaf nodes in the tree) subrelations. Then two sets of processes (one for leaf nodes and one for non-leaf nodes) are running in parallel. It is easy to verify that the index used for the *pipe* channels in the above program sets up the correct connections between the nodes. Since all the nodes (transputers) are running in parallel, full pipelining is achieved. That is, any non-leaf node is ready to consume any tuple sent by its child nodes at any time, thereby enhancing performance.

10.6 The future – the way forward

This chapter has briefly considered the role of parallel processing in database systems, and the use of the transputer in constructing database machines in particular. It has demonstrated how a query may be mapped onto a set of transputers and how some fundamental operations, such as join and sort, may be implemented in parallel. While these are essential to any parallel database system, we have provided here only the briefest sketch of a major and promising research area. There is a much wider scope for exploiting parallelism further in database systems, particularly in non-conventional applications. For example, in a **deductive database system**, not only the explicit data may be stored, but also implicit data (rules) may be represented. The evaluation of a user query in a deductive database is usually much more time-consuming than in a relational database system, and hence the utilization of parallel processing techniques is even more crucial. Development of object-oriented database systems seems to suggest the same importance for parallel processing. It should be pointed out, however, that achieving parallelism in these systems is much harder than in a relational database system. While work has been done in these areas, further research should be expected.

Acknowledgements

We should like to acknowledge the work of our colleagues at the University of Ulster all of them have been associated with projects involving parallelism related to database systems: D.A. Bell, C.M. Shapcott, F. F. Cai and D. Guthrie.

References

DeWitt, D.J. (1979) DIRECT – A Multiprocessor Organisation for Supporting Relational Database Management System. *IEEE Trans. on Computers*, pp. 395–406.

Kitsuregawa, M., Tanaka, H. and Tohru, M. (1984) Architecture and Performance of Relational Algebra Machine GRACE. In: *Proc. of Conf. on Parallel Processing*, R. M. Keller (ed), pp. 241–250. Washington: IEEE Computer Society Press.

Leung, C.H.C., Ghogomu, H.T. and Mannock, K.L. (1990) High Performance Relational Database Systems on Transputers. In: *Applications of Transputers*, vol.1, D.J. Pritchard and C.J. Scott (eds), pp. 430–436. Amsterdam: IOS Press.

Chapter 11

Experience of Using Transputers in a Robot Programming and Control System: A Case Study

J.D.M. McKeever and D.R.W. Holton

11.1 Introduction

This chapter is a case study on the use of transputers in the programming and control system for a robot manipulator. The topics described are: the implementation of kinematics modelling on transputers; the use of a transputer as a coprocessor in an IBM PC compatible computer system; the use of the BBC 1 MHz bus protocol to interface transputers to a robot servo controller; and the formal specification, design and implementation of a transputer-based robot programming system.

Some time ago a project was undertaken to build a Pascal-based software environment for the TQ MA2000 teaching robot; its objectives were to enable experience of robot software construction to be gained, and to produce case study material to support an undergraduate course in robotics (McKeever, 1988). The facilities provided in the robot software environment were influenced by those of PasRo (Blume and Jakob, 1985); the initial implementation used Turbo

Pascal on an IBM PC. The system has since been used as the basis of a series of experiments with the general theme of exploiting opportunities for concurrency in the design and implementation of robot control systems. In one set of experiments, transputers have been introduced at various places in the PC-based system; in the other set, formal specification and design techniques have been used to produce a system implemented on a network of transputers. The aims of the experiments have been twofold: to improve the execution efficiency of the robot system implementation, and to contribute to the understanding of robot software systems by applying techniques of software specification and construction developed in other application areas.

Work is still in progress: in this chapter the experience gained to date is described and some intentions for future work are outlined. The experiments discussed are:

- using transputers to speed up kinematics modelling,
- a transputer coprocessor in a PC-based robot control system,
- interfacing the MA2000 robot controller to a transputer network,
- a transputer-based robot programming system.

Note that this approach to using transputers in robot systems is not claimed to be definitive: several other groups have reported on work in this area, including Mirab and Gawthorp (1990) and Zomaya and Morris (1989).

11.2 Using transputers to speed up kinematics modelling

11.2.1 Role of kinematics modelling in robotics

The majority of industrial robots in use at present are programmed by teach programming (also known as online programming). This method requires an operator to guide the robot through the desired task, which may involve point-to-point operations such as pick-and-place of components, or continuous path operations such as seam welding or paint spraying. Teach programming is implemented in the robot controller by sampling and recording the positions of the robot joints at appropriate time intervals; later, as the recorded sequence is played back, the robot joints return to the sampled positions and so the robot repeats the taught task. The method suffers from a number of disadvantages: accurate teaching can be slow and tedious; the robot is non-productive during the programming phase; if there are changes in the workpiece the taught programs are cumbersome to edit. A further disadvantage arises if the robot needs to interact with other objects or machines: teach programming software systems model only the robot joint space, and do not contain any representation of the real world Cartesian space in which other components exist.

Offline robot programming seeks to overcome the disadvantages mentioned above. In an offline robot programming environment the robot, its workplace and the objects to be manipulated are modelled mathematically. The robot task is described using a suitable notation (an offline programming language) based on this mathematical model, and the offline programming system then translates the task operations into movements of the robot joints which can be executed by the robot servo controller.

As a minimum requirement, the mathematical model in an offline programming system must be capable of describing the locations of objects in space. A body in 3D space has six degrees of freedom, that is, it can move in six independent ways: if the body is described with respect to a Cartesian coordinate system then any movement of the body is equivalent to rotation about one or more of the x, y and z axes (change of orientation) and/or translation along one or more of the axes (change of position). Hence the location of a body may be specified by its position and orientation in a reference coordinate system; or, in other words, by the amount which a coordinate system embedded in the object has been transformed by translation and rotation from the reference coordinate system. There are several commonly used formulations of position and orientation; in modelling for robotics it is convenient to use homogeneous transforms, which combine descriptions of position and orientation in a single entity. A homogeneous transform, or frame, is a 4×4 matrix: the first three column vectors of a frame (which are mutually orthogonal) describe the directions of the axes of a new coordinate system relative to the reference coordinate system; the fourth column vector describes the translation of its origin. (Homogeneous transforms also allow representation of scaling, for example, for computer graphics modelling; the scale factor is the fourth component of each homogeneous vector, and in robotics calculations it may be left implicit.)

Hence frames may be represented as follows:

```
type
   Vector = record x, y, z { w implicit}: real end;
   Frame = record t, o, a, p : Vector end;
```

A kinematics model relates the position and orientation of a robot manipulator's end effector (for instance, gripper, spray gun, etc.) to its joint angles. When the kinematics model for a particular robot has been constructed, the position and orientation of the end effector corresponding to a given set of joint angles (or joint vector) may be determined; this is known as **direct kinematics**. Solving the **inverse kinematics** problem (that is, finding a joint vector which will result in a given end effector position and orientation) is more computationally intensive than solving the direct kinematics problem, and in general there are multiple solutions (just as the human hand can be placed in a desired position and orientation in several ways, for example with elbow up or down). The need for execution efficiency in robot kinematics modelling is important for many offline robot programming applications. For example, the robot may be required to track another moving object, or path control may be required, where the end effector must follow a predefined path in space. Both of these involve computing a sequence of end effector configurations in Cartesian space then solving the inverse kinematics problem for each configuration to determine the corresponding set of angles in joint space.

The goal of this experiment is to achieve acceptably efficient kinematics modelling by the use of concurrency.

11.2.2 Kinematics model of MA2000

The kinematics model of the MA2000 was developed using the technique described by Paul (1981). The algorithm to solve the inverse kinematics problem first finds all possible sets of joint angles for the required end effector position (there may be up to eight such sets), then selects the joint vector which will require least movement of the joints from their previous configuration. A skeleton outline of the algorithm coded in Pascal is described below:

```
type
    Vector = record x, y, z { w implicit}: real end;
    Frame = record t, o, a, p : Vector end;
    Angle = real;
    aJointVector = array [1..6] of Angle;

procedure InverseKinematics     (TheFrame : Frame;
                                 GivenJoints : aJointVector;
                                 var SolutionVector : aJointVector;
                                 var ErrorsFound : ErrorBitSet);

    type
        JointConfigurations =    record
                                     N : 0..8;
                                     Config : array [1..8] of aJointVector
                                 end;

    var
        JointConfig : JointConfigurations;
        I : Integer;

    begin { InverseKinematics }
        AllInverseKinematics (TheFrame, GivenJoints, JointConfig,
        ErrorsFound);
        if ErrorsFound = NoError then
            ChooseNearestConfiguration (GivenJoints, JointConfig,
            SolutionVector);
     end { InverseKinematics};
```

The AllInverseKinematics algorithm proceeds by finding the two possible values for the first angle, θ_1; each of these values is used to find two possible

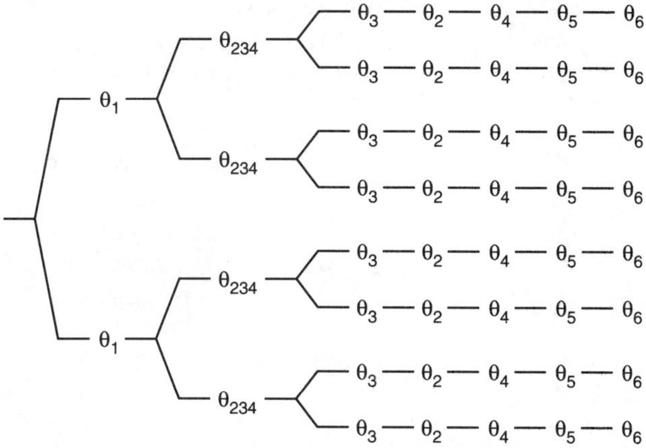

Figure 11.1 Representation of the AllInverseKinematics algorithm.

values for the sum of the next three angles, θ_{234}; each of these values is used in turn to find two possible values for the third angle, θ_3 which is used to find a set of values for the remaining angles, θ_2, θ_4, θ_5 and θ_6. Figure 11.1 illustrates the order in which the angles are found.

The solution as it is computed is built up in the record Configuration which is defined as a tree structure whose nodes store the values of angles (and any errors found). When all angle values have been found, the configuration tree is traversed and all sets of solution joint angles which are error-free are extracted into the output parameter SolutionConfig. (Note that a degenerate case can occur in which any value may be chosen for the angle θ_{234}; for brevity the details of this are omitted from the present description.)

```
procedure AllInverseKinematics    (TheFrame : Frame;
                                   GivenJoints : aJointVector;
                                   var SolutionConfig : JointConfigurations;
                                   var ErrorsFound : ErrorBitSet);

    { Finds all solutions for the given frame }

var
        Configuration :
        record
            Error : ErrorBitSet { TooClose };
                Configuration1 : array [1..2] of { ie two values of Theta1 }
                    record
                        Error1 : ErrorBitSet;
                        Theta1 {–3Pi/4 .. 3Pi/4} : Angle;
                            case State234 : (NonDegenerate, Degenerate) of
```

```
                    NonDegenerate :
                      (Configuration234 : array [1..2] of
                        {ie two values of Theta234 }
                          record
                            Error234 : ErrorBitSet;
                            Theta234 {–Pi/2 .. Pi/2},
                            Theta5   {–Pi/2 .. Pi/2},
                            Theta6   {–Pi/2 .. Pi/2} : Angle;
                            Configuration3 : array [1..2] of
                              {ie two values of Theta3}
                                record
                                  Error3 : ErrorBitSet;
                                  Theta3 {–3Pi/4 .. 3Pi/4},
                                  Theta2 {–3Pi/4 .. 3Pi/4},
                                  Theta4 {–Pi/2  .. Pi/2}  :
                                  Angle
                                end
                          end);
                    Degenerate : (...)
              end
      end;
  begin ... end { AllInverseKinematics };
```

Joint angle values are usually evaluated in terms of ATan2, the two parameter arctan function; the parameters to this are typically expressions involving sine and cosine of other joint angles. For example, the calculation of the composite angle Theta234 expands to:

```
C1 := Cos (Theta1);
S1 := Sin (Theta1);
Iota := a.x*C1 + a.y*S1;
Theta234 := ATan2 (–Iota, a.z);
```

(Common subexpressions which will be required again are assigned to the auxiliary variables C1, S1 and Iota to save recalculation.)

The initial implementation of the kinematics model used Turbo Pascal Version 5. Execution times for a typical inverse kinematic calculation generating four possible sets of joint angles were measured (see Table 11.1).

11.2.3 Inverse kinematics on transputers: Sequential version

In the first experiment with transputers the Turbo Pascal version of the inverse kinematics algorithm was converted directly to 3L Pascal and executed on a PC hosted B004 system containing a T800 transputer. It is perhaps noteworthy

Table 11.1 Time taken to execute the inverse kinematics algorithm on various hardware.

Machine (processor)	*Representation of reals (bytes)*	*Time (ms)*
IBM PC AT (Software floating point)	6	157
IBM PC AT (80287 Numeric coprocessor)	4	34
IBM PC AT (T800 Transputer)	4	6

that the Pascal implementation proved to be highly portable: the Turbo Pascal code was compiled on the transputer system with only trivial modifications. The results obtained are summarized in Table 11.1.

11.2.4 Exploiting concurrency

In subsequent experiments ways have been sought of exploiting the potential for concurrency in the inverse kinematics computation by distributing it over a network of transputers. One possible method for distributing the inverse kinematics algorithm is to activate two new processes to continue the computation as each pair of angles is computed. However, the current state of the computation must be communicated to each new process which is activated. To reduce this communication overhead, two techniques suggest themselves: a 'forking' technique, where just one additional process is engaged as soon as the first angle in each pair has been computed (Figure 11.3); and a 'Hungarian Army' technique, where the required end effector configuration is broadcast to eight processes which embark independently, each dedicated to following a path towards one potential solution (Figure 11.2).

The Hungarian Army algorithm exploits the tree structure of the solution. Eight independent processes are introduced, one for each path through the kinematics solution. Each process is given the Cartesian point to transform and a 'destination address' in the range 0–7 to specify which particular path through the computation it is to follow. The three digits of the binary representation of the 'destination address' indicate which branch the process is to take at each of the three nodes it encounters: take left branch if 0; right branch if 1 (Figure 11.2). For example, the process given destination address 4 (100 binary) will take the path to compute the set of joint angles labelled P_4.

Note that the technique of using the digits of the destination address to control routing may be generalized to other algorithms which have a tree structure: if the degree of the nodes at each stage of the algorithm is *b* then treat the

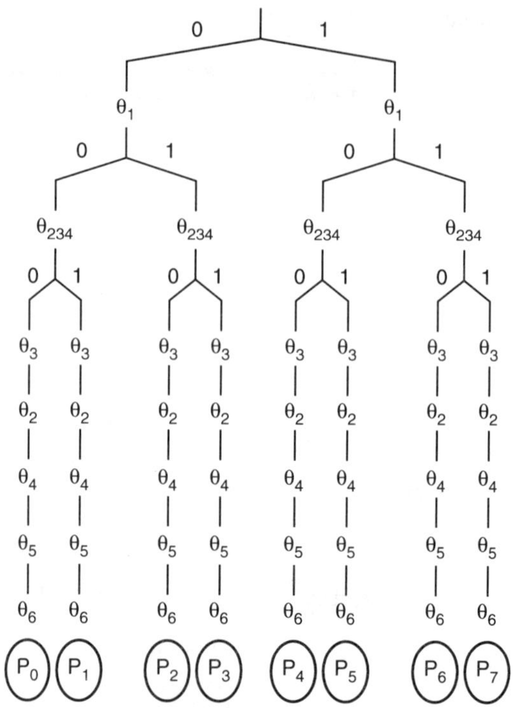

Figure 11.2 Diagrammatic representation of the Hungarian Army algorithm.

address as a base *b* number. (In fact, the degree of nodes at one stage need not necessarily be equal to the degree of nodes at the other stages: in this case treat the address as a mixed base number.)

In the forking algorithm process P_n starts up process P_{n+2^i} where $i = 0$, 1, 2. The first process, P_0, receives the configuration of the end effector which is to be translated to joint angles. It calculates both values of the first joint angle and passes the first one to process P_1, retaining the second. P_0 and P_1 each calculate two of the four values of the sum of the next three angles in parallel. P_0 passes one value to P_2 and retains the other value, while P_1 passes one value to P_3. P_0, P_1, P_2 and P_3 each calculate the two values of θ_3, pass one value to P_4, P_5, P_6 and P_7, respectively, and all eight processes calculate the remaining joint angles in parallel.

A version of the Hungarian Army algorithm has been constructed where each process is implemented as a task in a processor farm, using the 3L farm software (3L, 1989) for convenience. The performance achieved so far is disappointing – see Table 11.2 for the results obtained. Note that the sequential version of the inverse kinematics algorithm performed significantly better than the Hungarian Army algorithm, even when it is distributed. Part of the problem is that the implementor has no knowledge of, or control over, farm management communications overheads. It is reasoned that these overheads are large

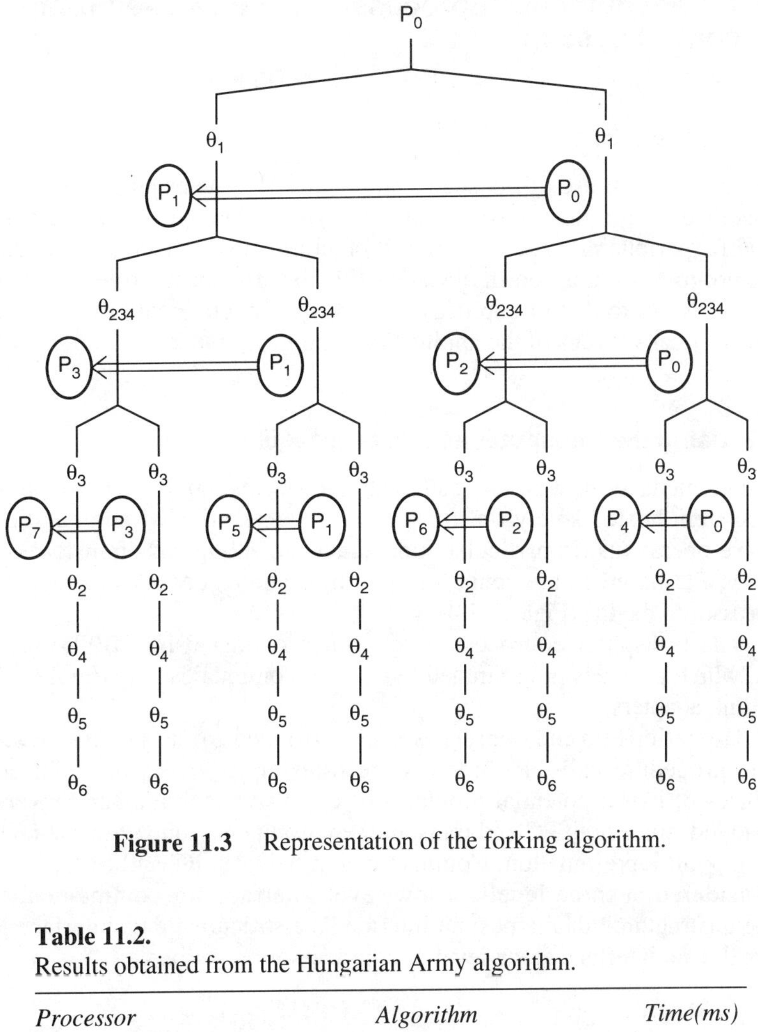

Figure 11.3 Representation of the forking algorithm.

Table 11.2.
Results obtained from the Hungarian Army algorithm.

Processor	*Algorithm*	*Time(ms)*
Single transputer	Sequential	6
Single transputer	Hungarian	21
Four transputers	Hungarian	11

because the farm management software is provided as a generic solution to implementing processor farms, and is therefore intended to be as general as possible. It is not expected that it would be able to perform as well as a custom written application, so it is planned to write a dedicated version with one transputer for each process and explicit communication, bypassing the 3L farm management software. However, 3L are about to offer a broadcasting operation for processor farms and this may turn out to be more efficient.

11.3 A transputer as coprocessor in a PC-based robot control system

11.3.1 Introduction

In the conventional use of transputer boards in PC systems an application program is loaded onto the transputer network, which then becomes master of the PC and its peripherals; access to the PC facilities is provided via a dedicated server program running on the PC. For this experiment the role of the transputer is changed to that of coprocessor, and its computational power is used to execute critical sections of the application running on the PC.

11.3.2 Using the transputer as a coprocessor

As an intermediate stage on the road to an implementation of the robot programming system hosted solely on transputers, the execution of computationally intensive operations (in particular, kinematics modelling) was transferred from the IBM PC-based implementation to a transputer network connected as a coprocessor to the PC (Figure 11.4).

The transputer coprocessor network was hosted in a B004 interface board, which provides programmed byte serial communication with the PC via C012 link adapters.

Operations on the coprocessor are provided to the PC effectively as remote procedure calls. Parameters are transferred across the interface as sequences of bytes; potential problems of expensive real number conversions are avoided, since both PC and transputer environments support the same IEEE floating point representation. Communication between PC and transputer may be considered at three levels: a low-level interface for communication of simple unstructured data types; an intermediate structured data interface level; and finally, an interface at the level of coprocessor operations.

11.3.3 Low-level interface operations

At the lowest level, routines are provided for transparent communication of bytes, 32-bit words and general sequences of bytes between the PC and transputer. This initial version of the low level interface operates synchronously, that is the PC is executing a busy wait operation (testing the C012 receive status register) while the transputer is performing the kinematics computations.

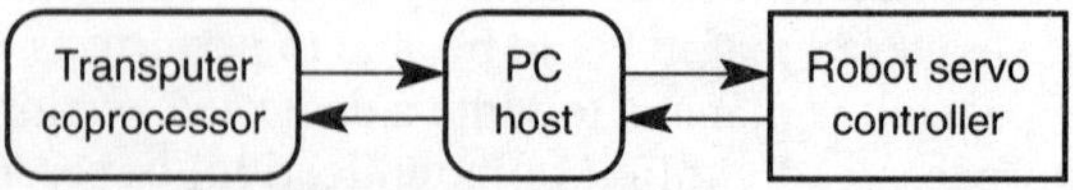

Figure 11.4 System with transputer coprocessor.

PC side of low level interface

```
const
    { ===== B004 Registers (mapped in PC Port address space) ===== }
    InputDataReg   = $150;
    InputStatusReg  = $152;
    OutputDataReg   = $151;
    OutputStatusReg = $153;

    { ===== B004 status register bit values ===== }
    InputReady     = $01;
    OutputReady     = $01;


{ simplified for purposes of exposition: assumes response from links }

procedure SendByte (TheByte:Byte);
  begin
    while (Port [OutputStatusReg] AND OutputReady) <> OutputReady do
      {idle };
      Port [OutputDataReg] := TheByte;
  end;

procedure ReceiveByte (var TheByte : Byte);
  begin
    while ((Port [InputStatusReg] AND InputReady) <> InputReady) do
      { idle };
      TheByte := Port [InputDataReg];
  end;
```

Four further routines are provided: Send32BitWord and SendByteArray which are implemented as calls to SendByte, and Receive32BitWord and ReceiveByteArray which build upon ReceiveByte.

Transputer side of low level interface

The low level interface on the transputer side mirrors that on the PC side, but in this case the necessary routines are already available in the 3L Parallel Pascal library:

```
$INCLUDE 'chan.inc'

{ Include channel communication package from 3L Parallel Pascal library;}
{ provides operations:}
{       Chan_Out_Byte  Chan_Out_Word  Chan_Out_Message }
{       Chan_In_Byte   Chan_In_Word   Chan_In_Message  }
```

11.3.4 Structured data interface level

The interface protocol used to exchange structured data between PC and transputer is a subset of the AFServer protocol (3L, 1989) used in PC-based transputer development systems. This is a two-level tagged protocol: at the lower level each piece of information transmitted is prefixed by a tag byte which identifies it as a byte, word, record, etc.; at the higher level, tag values identify operations requested of the server. Although the AFServer protocol is somewhat over-elaborate for this application, there appeared to be little benefit in developing an alternative (simpler) protocol; in addition, limitations in an early release of the system software used in the initial experiments meant that the transputer side of the interface had to use a filter task which adhered to the AFserver protocol, hence the use of that protocol was unavoidable.

PC side of structured data interface

On the PC, two operations are provided for communication between the PC and transputer, GetBlockOfBytes and PutBlockOfBytes:

```
const
    Nilrecord_Value = 8;
    Record32_Value  = 12;

procedure PutBlockOfBytes (var B {: AnyBlockOfBytes};
                                 Size : Integer);

    begin
        if Size > 0 then
            begin
            SendByte (Record32_Value);
            Send32BitWord (Size);
            SendByteArray (B, Size)
            end
        else
            SendByte (NilRecord_Value);
    end;

procedure GetBlockOfBytes (var Count : Integer;
                                 var B {: AnyBlockOfBytes});

    var
        Tag : Byte;

    begin
        ReceiveByte (Tag);
        if Tag = Record32_Value then
```

```
            begin
            Receive32BitWord (Count);
            ReceiveByteArray (Count, B)
            end
        else
            begin
            TagError (Tag, Record32_Value);
            Count := 0
            end
    end;
```

Transputer side of structured data interface

```
const
    PCChan = 0;

procedure PutBlockOfBytes (var B : univ Char;
                               Size : Integer);
    begin
        if Size > 0 then
            begin
            Chan_Out_Byte (Chr (Record32_Value), Out_Chan^[PCChan]^);
            Chan_Out_Word (Size, Out_Chan^[PCChan]^);
            Chan_Out_Message (Size, B, Out_Chan^[PCChan]^) ;
            end
        else
            Chan_Out_Byte (Chr (NilRecord_Value), Out_Chan^[PCChan]^);
    end;

procedure GetBlockOfBytes (var Count : Integer;
                           var B : univ Char);

    var
        Tag : Char;

    begin
        Chan_In_Byte (Tag, In_Chan^[PCChan]^);
        if Tag = Chr (Record32_Value) then
            begin
            Chan_In_Word (Count, In_Chan^[PCChan]^);
            Chan_In_Message (Count, B, In_Chan^[PCChan]^)
            end
        else
            Count := 0
    end;
```

11.3.5 Coprocessor interface level

Finally, the operations on the coprocessor are considered.

PC side of coprocessor interface

The PC interface module first initializes the transputer network by loading a code file, then makes available the operation DoKinematicsOnTransputer.

```
procedure DoKinematicsOnTransputer (var CurrentJoints : aJointVector;
                                    var SolutionVector : aJointVector;
                                    var F : frame;
                                    var ErrorsFound : ErrorBitSet);

    var
        OutRec:   record
                      OutputFrame : Frame ;
                      JointsforOutput : aJointVector
                  end;

        InRec: record
                  JointsInput : aJointVector;
                  ErrorsInput : ErrorBitSet
               end;

        Count : Integer;

    begin
        OutRec.OutputFrame := F;
        OutRec.JointsforOutput := CurrentJoints;
        PutBlockOfBytes (OutRec, SizeOf (OutRec));

        { Transputer now performs inverse kinematics }

        GetBlockOfBytes (Count, InRec);
        if Count = SizeOf (InRec) then
            begin
            SolutionVector:= InRec.JointsInput;
            ErrorsFound := InRec.ErrorsInput;
            end
        else
            { some appropriate error action }
    end { DoKinematicsOnTransputer };
```

Transputer side of coprocessor interface

The transputer executes in an infinite loop, accepting requests for kinematics conversions, performing the conversion and returning the results:

```
static InRec:  record
                  InputFrame : Frame;
                  InputJoints : aJointVector
               end;

static OutRec:record
                  SolutionOutput : aJointVector;
                  ErrorsOutput  : ErrorBitSet
               end;

   var
      Count : Integer;

   begin
      while true do
         begin
         GetBlockOfBytes (Count, InRec);
         if Count <> size(InRec) then { error };

         InverseKinematics    (InRec.InputFrame,InRec.InputJoints,
                               OutRec.SolutionOutput,
                               OutRec.Errors Output);

         { 3L Pascal rounds record sizes up to multiple of 4 bytes }
         PutBlockOfBytes (OutRec, size (aJointVector) +
         size (ErrorBitSet) );
         end
   end.
```

11.3.6 Coprocessor performance

This technique has proved successful for inverse kinematics evaluation, with a time of 12 ms required to perform a typical inverse kinematics transformation. Comparing this with the results obtained in Table 11.1, it can be seen that, even with the communication overheads between the processors, using the T800 as a coprocessor is better than a factor of two faster than using the 80287 numeric coprocessor, and is an order of magnitude faster than the best performance available from the PC AT without a coprocessor (that is, version with six-byte reals).

11.3.7 Possible extensions/enhancements

An asynchronous version of the system would enable possible concurrency between the PC host and the transputer to be exploited. For example, if path control as mentioned in Section 11.2 is considered, the PC host could calculate the next point in Cartesian space while the previous point is transformed into joint space, with the result from each transformation being collected before the next point to be transformed is transmitted.

It is planned to develop such an asynchronous version of the system, but this will require the replacement of the B004 board by a later revision B00x interface board which has PC interrupts available. It should also be possible to reduce the communications overhead by replacing the slow programmed transfers between PC and transputer (which constitute a major bottleneck in the system) with faster DMA transfers. A further possibility is to exploit the 16-bit string-mode transfers which are provided on TMB16 boards (Transtech, 1991); these use a T2 transputer memory-mapped into the PC input-output address space as a dedicated communications processor.

Although just one operation (inverse kinematics modelling) is implemented on the transputer, it would be trivially easy to extend the idea to provide a range of operations, with an initial tag value used to select among the available operations and to control the size of data transfer for parameters.

Note that the idea of a transputer as coprocessor is applicable to many systems hosted on PCs which have critical sections requiring fast processing.

11.4 Interfacing the MA2000 robot controller to a transputer network

11.4.1 Background

The next experiment had the goal of reducing the communication overheads mentioned in the previous section by interfacing the network of transputers directly to the robot servo controller. The arrangement is illustrated in Figure 11.5.

The MA2000 robot contains a PID controller implemented using 12-bit resolution digital to analogue and analogue to digital converters, and a 6502 processor. This controller was designed to interface to a BBC micro host computer acting as a slave device on the BBC 1 MHz bus (Bray *et al.*, 1983).

The design phase was split into two phases: designing the hardware to connect transputer links to the controller; and designing the software to handle the bus protocol.

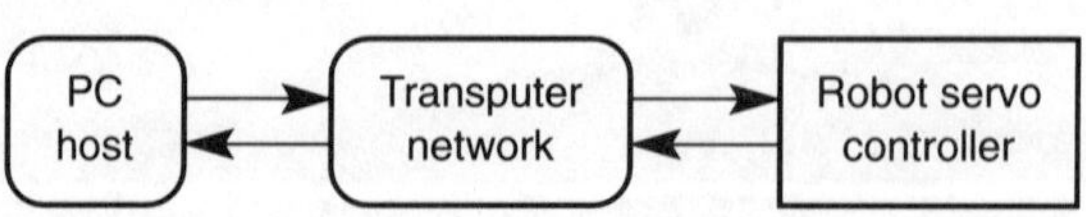

Figure 11.5 Interfacing transputers directly to the robot controller.

11.4.2 Hardware design

The robot controller has an area of shared memory and a user VIA port which are used to transfer information between the host computer and the controller. These are accessed by the host via the BBC 1 MHz bus. The problem is therefore to design hardware to enable a transputer to emulate a BBC 1 MHz bus controller.

Transputers communicate data in serial over hardware links, while the BBC 1 MHz bus requires parallel data, requiring the use of C011 link adaptors to convert the serial data stream to parallel data. The BBC 1 MHz bus protocol consists of three components: Address, Data and Control. Therefore, three C011 link adaptors were required to interface a transputer to the robot controller; one each for Address, Data and Control signals. A schematic diagram of the hardware interface is shown in Figure 11.6.

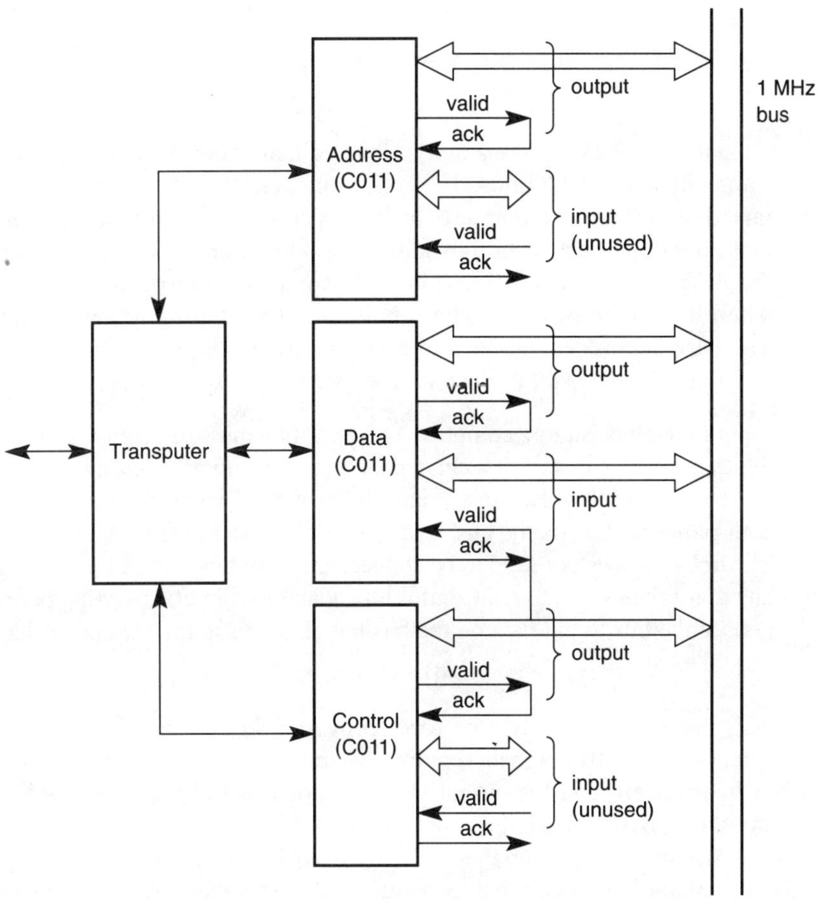

Figure 11.6 Hardware interface to the MA2000 controller.

Output is only required from the Address and Control components of the interface; this is because the transputer is the bus master and has no need to receive address information as it decides which memory address to read, and it sends the requisite control signals to peripherals on the bus. Data, however, has to be transferred both on and off the bus, so both the input and output sections of the Data link adaptor need to be connected to the bus.

The C011 link adaptor uses a standard two wire handshake to validate data transmission. One line, labelled *valid* in Figure 11.6, is taken high by the transmitting device to indicate transmission of a 'valid' data stream. The other line, labelled *ack*, is taken high by the receiving device to acknowledge the receipt of the data stream. The transmitter takes *valid* low again when it has finished transmitting, and the handshaking is completed by the receiver taking *ack* low when it has received the data stream.

However, the BBC 1 MHz bus has no signals which could be used to perform this handshaking, so the *valid* and *ack* signals must come from the hardware interface.

The hardware interface must account for both input and output to the 1 MHz bus:

- Output to the bus is more simply achieved than input. It is clear from the Inmos data sheets (Inmos, 1989) that the acknowledge signal may be raised as soon as the output data stream commences; it is not required that all the data has been transmitted. It is therefore possible to connect the *valid* line to the *ack* line (Figure 11.6); as the transputer raises *valid* when it starts to transmit data it is in effect validating its own output. The peripheral does not need to perform any handshaking, and the data remains on the bus until it is replaced by the next item of data.

- Input to the bus is more complex. The *valid* line must be taken high before the transputer will start to receive data, but there are no connections to the bus to achieve this. However, only three of the control lines are used to control peripherals on the bus, so a spare one is used to raise the *valid* line. The *ack* line does not need to be connected to anything. Therefore, sending an appropriate value to the Control link adaptor will not only tell a peripheral on the bus to put data on the bus but also inform the transputer that it should expect a data value.

This interface is merely a protocol converter, in this case from serial to parallel data. The control signals required by the link adaptors come from the interface board itself, and the output from the board is indistinguishable from the output a real BBC 1 MHz bus controller would provide. A further point to note is that the synchronization that needs to occur between the robot controller and the transputer is software based, ensuring that this interface is completely general and can be used to control any peripheral device on a BBC 1 MHz bus.

11.4.3 Software design

The design of the software has been split into two tasks: the hardware interface driver and the higher level interface to the host software. Figure 11.7 is a schematic diagram of the structure of the software.

Hardware interface driver

The hardware interface driver task must emulate the master of a BBC 1 MHz bus, that is it addresses a device through the interface rather than being addressed by the device. As noted above, the master requires both input and output for the data stream alone; the Address and Control streams are output only. The hardware driver must interface to two distinct components of the robot controller: the shared memory and the user Versatile Interface Adaptor (VIA). The shared memory is an area where data may be exchanged with the robot controller. The VIA consists of two ports, port A and port B, each of which has a read only and a write only register, and provides access to several control functions provided by the robot controller such as synchronizing access to the shared memory.

The hardware interface driver is only required to provide facilities for reading and writing to the shared memory and both ports of the VIA, and to initialize the interface before allowing data to be read from or written to the robot controller. This ensures that this task is very simple and contains no code which is specific to this application; it could be used to control any peripheral on a BBC 1 MHz bus with similar facilities.

Software interface

This task is required to provide higher level facilities to the control software than those provided by the hardware interface driver, that is it must allow easy access to the data provided by the robot controller. It makes use of the facilities for reading and writing to the robot controller provided by the hardware interface driver, and is therefore not dependent upon the particular method used to access the shared memory or the VIA.

This interface must provide equivalent operations to those that were available on the PC host computer so that the control software can be moved to a transputer with minimum effort. The operations that are required include

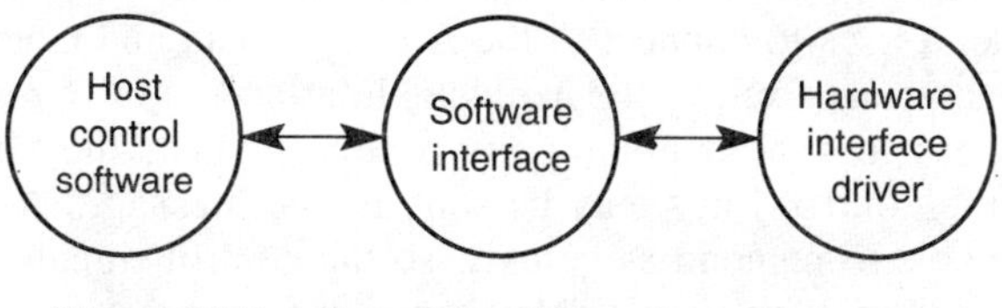

Figure 11.7 Overview of the software structure.

acquiring access to the shared memory, reading the current position of the robot and setting the state of the end effector.

11.4.4 Software implementation

This section describes the implementation of the hardware and software interfaces, though the hardware interface driver will be examined in more detail.

Hardware interface driver implementation

The implementation of the operations to read from the robot controller are given and explained below:

```
procedure Get ; {To obtain a byte from shared memory}

    begin
        chan_in_byte (address, in_chan^[0]^);
        chan_out_byte (address, Link1Output^);
        chan_out_message (3, read_Valid_reset, Link3Output^);
        chan_in_byte (data, Link2Input^);
        chan_out_byte (data, out_chan^[0]^);
    end {Get};
```

This procedure is used to read a byte from shared memory. It receives the address of the byte to read from the software interface task via logical input channel zero. This address is then sent to link 1, which is attached to the Address link adaptor (Figure 11.6). The address is therefore placed on the 1 MHz bus. Next a message is sent to link 3 which is attached to the Control link adaptor. This message is made up of three bytes: the first tells the controller's shared memory that a read operation is about to occur; the second is used to validate the data from the shared memory to the Data link adaptor; and the third byte tells the shared memory that the operation was successful and readies it for the next operation. The data from the shared memory is then read from link 2 and sent to the software interface task via logical output channel zero.

It is worth expanding a couple of points from the description above. Logical input and output channels are used to pass data between the hardware interface driver task and the software interface for the sake of flexibility; this allows the the decision as to whether these tasks should be placed on the same transputer or a different one to be deferred until configuring the complete application. Conversely, the hardware interface task must explicitly name the hardware links it uses to connect to the Address, Data and Control link adaptors, because these are fixed by the hardware interface.

As access to shared memory is relatively fast, the signals that are required on the control lines may be sent as one message. A slower device would not be able to respond as quickly, so the control signals would have to be sent one at a time, with a short delay between each one.

The procedure to write to shared memory is very similar to the procedure to read from shared memory, so will not be presented. The procedure for reading a byte from the VIA is presented next:

```
procedure Get_Port ; {To obtain a byte of data from the user VIA port}

    begin
        chan_in_byte (address, in_chan^[0]^);
        chan_out_byte (address, Link1Output^);
        chan_out_byte (read_byte, Link3output^);
        chan_out_byte (valid_byte, Link3output^);
        chan_out_byte (reset_byte, Link3output^);
        chan_in_byte (data, Link2Input^);
        chan_out_byte (data, out_chan^[0]^);
    end {Get_Port};
```

This procedure has many similarities with the previous one. It also receives an address, in this case either of port A or port B, and sends it to the Address link adaptor. This time, however, the control signals are sent one at a time because the VIA is a much slower device than the shared memory. The overheads involved with calling the chan_out_byte procedure and de-referencing the pointer to the link provide a long enough delay for the VIA to respond. Once the control messages have been sent the data is read and transferred to the software interface.

Again, the procedure to write to the VIA is very similar to this procedure so will not be presented.

The main body of this task first initializes the interface, then repeatedly waits for a request from the software controller and calls the appropriate procedure to deal with the request until a message telling it to terminate is received:

```
begin
    Initialize_Interface;
    repeat
        {Wait for software controller synchronization signal}
        chan_in_byte (choice, in_chan^[0]^);
        case choice of
            to_read         :       get;
            to_write        :       put;
            read_port       :       get_port;
            write_port      :       put_port;
            to_terminate    :
        end;
    until choice = to_terminate; {HALT condition}
end. {Hardware Driver Interface}
```

Software interface implementation

The majority of the services offered in this task are specific to the robot controller device, and so are not particularly relevant here. One procedure is given as an example of how the facilities of the hardware interface driver are used, and a larger example is given to show one of the services offered by this task to the host control software:

```
procedure WriteInterface (location : abyte ; value : abyte);

    begin
        chan_out_byte (to_write, out_chan^[2]^);
        chan_out_byte (location, out_chan^[2]^);
        chan_out_byte (value, out_chan^[2]^);
    end {WriteInterface};
```

This procedure is very simple; it first sends a message to the hardware interface to tell it to write a byte to the interface. The address of the byte is then sent, followed by the value to be written.

The following is an example of a higher level facility that is available to the host control software:

```
function ReadASetPoint (Joint : aJoint) : aSetpoint;

    {Read a setpoint of a particular joint}

    var setpoint : aSetpoint;

    begin
        AcquireSharedMemory (true);
        setpoint := ReadInterface (positionshibase + joint – i);
        setpoint := setpoint << 8;
        setpoint := setpoint + ReadInterface (positionslobase + joint – i);
        ReleaseSharedMemory;
        ReadASetPoint := setpoint;
    end {ReadASetPoint};
```

A setpoint is a 12-bit number indicating the current position of a particular joint. This procedure first acquires exclusive access to the shared memory; the parameter true in the procedure call is used to indicate that access to the memory should be acquired after the controller has accessed it, so the most up-to-date value of the setpoint may be read. As a setpoint is 12 bits, the shared memory needs to be accessed twice, once to read the high part of the position and again to read the low part. The high part is stored in a temporary variable and is shifted left eight places. The low part is then read and added to this giving the actual reading. Once the memory has been read exclusive access to it is released.

11.4.5 Evaluation

An interface to the MA2000 robot controller has been designed and implemented successfully. The advantage of this approach was that it avoided the bottleneck of communication which existed when using the transputers as a coprocessor. It had been found that there was a significant overhead incurred in passing data between the PC host and the transputer network. In tests, using this interface gave a 42% decrease in the amount of time it took to transfer a precalculated path to the robot controller when compared to the case where the points had to be transferred to the PC host and then transmitted to the robot controller.

The modular nature of the design has meant that the general purpose hardware interface and the software to drive it can be used as a BBC 1 MHz bus emulator in many different circumstances. The addition of a software interface task to this hardware interface has eased the burden in the host control software of communicating with the robot controller.

11.5 A transputer-based robot programming system

11.5.1 Introduction

A system for facilitating offline programming of robots has been designed, specified formally and implemented on a network of transputers (Holton, 1991). This system is akin to a conventional operating system as it is an interface between the hardware and the user, hiding most of the implementation details behind a set of primitive operations that allow the robot to be programmed easily.

11.5.2 Design

The initial part of the design was concerned with enumerating the primitive operations required by the system. A fairly basic set of primitives was chosen, including:

- motion control in Cartesian space,
- motion control in joint space,
- relative motion in Cartesian space,
- report current position (in Cartesian or joint space), and
- move to new position in joint space.

Motion control is important because the fundamental use of a robot is to move workpieces from one place to another. It was decided to allow motion control in both Cartesian and joint space because these operations are complementary; some applications may only require movements in joint space, typically where the actual path the robot takes does not need to be constrained. Other applications may require the robot arm's tool tip to follow a particular path through space – an example of this is spray painting. It is far easier to define such paths in Cartesian space.

Relative motion in Cartesian space is important in many applications, particularly when external sensors are used with a robot arm to guide it to a required position. For example, a robot arm may be guided to a workpiece by viewing a scene which incorporates the end effector and the workpiece, calculating a motion which will tend to decrease the distance between them, and repeating these actions until the workpiece can be picked up. The vector calculated in this case is a relative movement; there is no knowledge of the absolute position of the sensor, the end-effector or the workpiece.

In some circumstances, it is important that the application program is provided with the actual position of the manipulator, rather than assuming that it is where it was last commanded to move. The primitives for reporting the current position in both Cartesian and joint space provide the means for giving feedback on the actual position of the manipulator. A good example of where this might be used is error detection; one obvious error is that the robot arm is not actually in the position it is supposed to be, and if there is a gross difference between the desired and actual position it could be that the manipulator has collided with something.

Similarly, it is important to provide the experienced programmer with the means to move the robot directly to a particular position, hence the provision of the primitive to move the robot to a new position in joint space. This could be used to initialize the robot in a certain position, or to implement operations which had been left out of the primitives, such as a force guarded motion.

These primitives were chosen because it was believed that they would provide a programmer with the facilities required to implement quickly and easily application programs to solve the majority of tasks.

Once it had been decided which primitives to incorporate, an overall structure for the system was required. Here the design was heavily influenced by techniques advocated in the design of operating systems, that is, a layered structure with each layer building on the facilities provided by the previous one. This allows operations which are used in the implementation of the services provided by each layer to be effectively hidden from other layers, greatly simplifying the implementation, maintainence and use of each layer. The implementor is free to implement the services in the most appropriate way, as long as the interface to those services is consistent. The code can be more easily maintained because the effect of modifying much of the software is localized. The user is probably benefited most by this, as the only knowledge required of each layer is the services provided and how to use them correctly.

The choice of the layers incorporated in the system was dictated by a number of factors. The main consideration was that each layer should provide a more abstract robot arm to the layers which used its facilities. Another important consideration was to choose layers that corresponded to the layers that were apparently present in the physical system. A further consideration was the distribution amongst the layers of the primitives identified above. In general, it was attempted to ensure that operations in one layer took an order of magnitude less time to execute than those in the layer above (Figure 11.8).

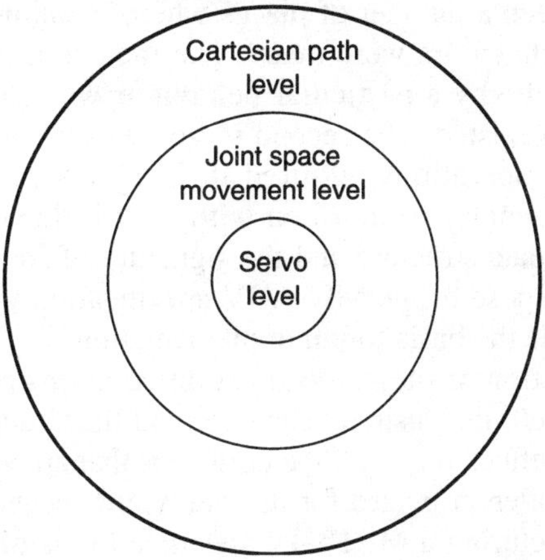

Figure 11.8 Structure of the robot programming system.

11.5.3 Specification

A formal specification of the design outlined above was constructed. The structure of the system was specified in LOTOS, a formal description language that is best known for its use in the description of Open Systems Interconnection (OSI) protocols. LOTOS consists of a process algebra similar to CCS and CSP, and a data typing mechanism based on ACT-ONE. LOTOS allows a system to be specified in terms of a collection of communicating processes composed sequentially or in parallel. This allowed the levels that were identified in the design phase to be explicitly represented, giving a clean separation of concerns in the specification.

The decision to use LOTOS was a pragmatic one, mostly influenced by the fact that some work was being done with LOTOS by colleagues and a tool for checking the syntax and semantics of a specification was available. However, it soon became clear that the data typing mechanism of LOTOS did not have the required expressive power; the operations that had to be specified were characterized by simple data structures and complicated operations upon them. The data typing mechanism in LOTOS is not well suited to this, so it was decided to use 'place-holders' in the LOTOS specification and specify the data structures and operations in META-IV, the specification language of VDM (Jones, 1990). META-IV is a model-oriented specification language which has a richer set of basic operations than ACT-ONE, making it far easier to specify the desired operations. The place-holders in the LOTOS specification were required for two reasons. The first was to provide an indication in the specification of the data types that were required, showing the types of values that were used to resolve choices between different behaviours. If this had not been present the LOTOS specification would have been less meaningful, as there would have been a number of places where seemingly non-deterministic choices between behaviours were made. The inclusion of data types made it easier to understand why a particular behaviour was chosen where choices between behaviours existed. The second reason was to provide a prototype for the data types and operations required in META-IV. The data types were specified in a rudimentary form, often with very little detail. In some cases, only the name of a data structure and the signature of some operations upon it were given. The purpose of the META-IV specification was to flesh out those prototypes to provide the basis for an implementation.

The combination of these specification notations proved to be a useful way of specifying both the desired behaviour and the structure of the system. It also seemed to deal effectively with the criticisms that are voiced about both languages. LOTOS is often criticized for its data typing mechanism, while the lack of facilities for structuring a META-IV specification is often cited as its main problem. Combining the two resulted in a specification that was well structured and had comprehensive and useful definitions of data structures and operations.

11.5.4 Implementation

The next step was to implement the specification (Figure 11.9). Until this time, little consideration had been given to the language and hardware environment that was to be used. It was decided to use 3L Parallel Pascal on a network of transputers. There were several reasons for choosing this environment:

- to ensure that the implementation exhibited the same structure as the specification;
- the translation of the large amount of sequential Pascal software which already existed for controlling the robot would have been time consuming and error prone;
- the implementation had the potential of meeting the time constraints of the problem; and
- the hardware and software were readily available.

The implementation proceeded informally, in that no formal proofs were constructed to show that the implementation actually satisfied the specification. Rules of thumb were formulated to aid the conversion of the LOTOS process algebra, and the algorithms which implemented the META-IV specifications were either derived from the specification, or it was shown that the particular algorithm satisfied its specification. It was noted that the META-IV data structures were in general easily represented in Pascal. Often a syntactic transformation was all that was required to construct a legal Pascal type definition from its META-IV specification.

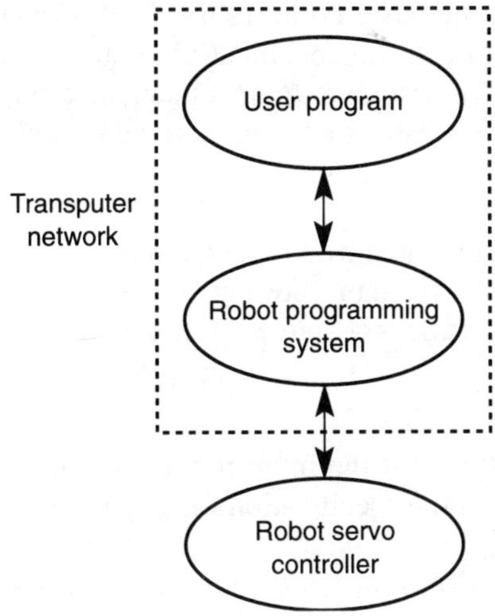

Figure 11.9 Implementation of the robot programming system.

The META-IV specifications were, in the most part, implemented with little difficulty. As mentioned above, in some cases it was possible to demonstrate that an implementation satisfied the specification, and in others it was possible to derive an implementation from its specification. This was principally because the data types were very simple, and the operations on those types were generally straightforward mathematical functions. In other circumstances, especially where the specified data structures do not correspond to ones available in the implementation language, the implementation may not progress as easily.

Implementing the LOTOS process algebra turned out to be the most problematic part of constructing the system. It has already been mentioned that the environment in which the finished program would run was not considered during the specification phase. This is considered to be good practice when formally specifying a system because it leads to the most general specification possible. This, of course, does not mean that it is the most easily implemented specification. The usual approach taken to the implementation of abstract, model oriented specifications is to make the specification more concrete so that it resembles the implementation language more closely. This can be done by refining the data structures, necessitating changes to the operations on those structures, refining the operations to make them more concrete, or applying transformations to the specification to produce an implementation. Common methods of refining specifications based on process algebras also include applying transformations to the specification; another method that is widely used is to construct a more concrete specification by hand and then prove that it is equivalent to the original specification. The approach taken in this project was to produce some rules of thumb which allowed the specification to be transformed to Parallel Pascal.

As noted above, the basic building block of a LOTOS specification is the process. As a result of this, a specification would incorporate many more processes than it would be desirable to implement, so a rule of thumb had to be devised which allowed processes to be transformed to procedures in certain circumstances. The rule is given below:

> If Process1 is instantiated by Process2, performs some actions then terminates, and Process2 can take part in no events until Process1 has terminated, then Process1 can be represented by a procedure in an implementation.

It was important to ensure that the interprocess communication in the implementation matched that of the specification for two reasons:

- if they exhibited different behaviours in this respect then it would be difficult to show, however informally, that the implementation satisfied the specification; and
- as there was a large degree of confidence that the specification was deadlock free, ensuring that the implementation exhibited the same communication behaviour as the specification would give the same degree of confidence to the implementation, providing that the calculations terminated.

A further problem arose when configuring the system. The specification had been constructed with many channels connecting the processes, which is encouraged in LOTOS because it adds to the clarity of the specification. However, transputers have four hardware links available for communication between transputers, though processes which share a particular transputer may use many more links to communicate. This meant that the actual distribution of processes over a network of transputers became an interesting problem of maximizing the utilization of each transputer while keeping the system within the constraints enforced by the transputer's hardware.

The main problem that was encountered throughout the implementation was the language used, 3L Parallel Pascal. It offered very little support to the programmer, for example the error checking facilities offered by the compiler were minimal. As each process is compiled on its own with no reference to the other processes it is to be combined with later on, it is very easy to make small, usually typographical, errors that have disastrous consequences. For example, it was possible to send data of one type from one process and receive it as a different type at another, an error that would have been picked up immediately if a language such as Ada had been used. When a program such as this was run it would perhaps work for a while and then stop, or perhaps hang up straight away with no indication of what was happening – very frustrating for the implementor.

A further source of frustration encountered was when configuring the processes as a single application. Again, sporadic errors were found; in one case an operation would succeed sometimes, but not always. When this was investigated more systematically it was found that for small sized problems it worked well, but when the size of the problem increased it failed. This indicated problems with the memory allocation for the process. When configured, that process should automatically have been given the remainder of the memory, using the facility for doing this in the configuration language. However, when an actual figure for the memory to be allocated to the process was added the system worked on all sizes of problems. This highlights another problem; the configuration section could only be derived in part from the specification of the system, hence leaving this crucial part of the program as the only one which was not subject to validation against an independent specification.

11.5.5 Evaluation

Despite the problems mentioned above with the implementation of the system, a working robot programming system was produced. It was used for a number of experiments to establish whether the initial design objectives were met, where it was established that the system could be used to support the programming of relatively complex tasks. It was also found that a distributed implementation had considerable improvements in runtime efficiency compared to the sequential system used previously, despite the fact that the most computationally intensive operations were implemented sequentially. The runtime benefits had been obtained just by pipelining operations, and it is expected that further improvements are readily obtainable.

The specification technique used proved to be successful. The sequential portions of the code, specified in META-IV, were readily transformed to the programming language. The implementation of the LOTOS process algebra was more problematic in some instances, but these problems were generally only apparent in the latter stages of the implementation when the hardware to be used was considered in more detail. An approach that could be taken to alleviate this problem is to produce the initial, abstract LOTOS specification, then transform it into an equivalent LOTOS specification which is more suitable for implementation on a network of transputers. LOTOS is well suited to this type of manipulation as there are rules which allow the equivalence of specifications to be proved. Such an approach would make LOTOS very suitable for the specification of software for transputer networks.

The major stumbling block encountered was the implementation language. 3L Parallel Pascal, because it is essentially a sequential programming language with extensions to allow parallel programs to be developed, was not well suited to the production of such a system. It is envisaged that a programming language specifically designed to support parallel programming would give far better support to the implementation of concurrent software generally, not only that which has been formally specified.

11.6 Conclusions

This chapter has described experiments in which transputers have been incorporated into a robot system in a number of ways: kinematics modelling has been implemented on transputers; a transputer has been used as coprocessor in a PC system; transputers have been interfaced to a robot servo controller; and a transputer-based robot programming system has been specified and implemented.

The experiments have succeeded in improving the performance of the robot system, and have contributed to the understanding of the design and implementation of such systems. Some work remains to be done: in particular, the implementation of kinematics modelling on transputers will benefit from further refinement. The performance of a transputer as coprocessor in a PC system is promising, and will be much improved if the overheads of programmed data transfer can be reduced. The interface to the robot servo controller may have more general application in linking transputers to devices which use the BBC 1 MHz bus protocol. Despite the problems resulting from the choice of implementation language, the experience of producing a formal specification of a robot programming system and of implementing it on transputers has been worthwhile.

Acknowledgements

It is a pleasure to thank Tony Donnelly, Maciej Czyrny and Thomas McManus for their contributions to the kinematics implementations; John Flanigan and Fergus Wills for their work on the interface to the MA2000 PID controller; the SERC and TecQuipment International for supporting DRWH on a CASE studentship; and the Northern Ireland Transputer Centre for equipment and assistance.

References

3L (1989) *Parallel Pascal User Guide*. 3L Ltd.

Blume, C. and Jakob, W. (1985) *PasRo: Pascal for Robots*. Springer-Verlag, Berlin.

Bray, A.C., Dickens, A.C. and Holmes, M.A. (1983) *The Advanced User Guide for the BBC Micro*. Cambridge Microcomputer Centre, Cambridge, UK.

Holton, D.R.W. (1991) *A Rigorous Approach to Robot Programming*. PhD thesis, The Queen's University of Belfast, Northern Ireland.

Inmos Ltd. (1989) *The Transputer Data Book*. Inmos Ltd. London, UK.

Jones, C.B. (1990) *Sytematic Software Development Using VDM*. Prentice-Hall International.

McKeever, J.D.M. (1988) Introductory robotics for computer science students. *University Computing*, **10**, pp.194–199.

Mirab, H. and Gawthrop, P.J. (1990) *Transputers for Robot Control*. Technical Report, Department of Mechanical Engineering, The University, Glasgow.

Paul, R.P. (1981) *Robot Manipulators: Mathematics, Programming and Control*. MIT Press, Cambridge, MA.

Transtech (1991) *TMB16 Hardware: 16-bit PC TRAM board*. Transtech Parallel Technology.

Zomaya, A.Y. and Morris, A.S. (1989) Fast robot kinematic modelling via transputer networks. In: *Parallel Processing and Artificial Intelligence* (Reeve, M. and Zenith, S.E., eds.). Wiley, Chichester, UK.

Chapter 12
Ray Traced Graphics

R.M. McKeag

12.1 Introduction

This chapter has two purposes: it describes how an existing sequential (Pascal) program has been ported to a transputer and parallelized to run on a processor farm; and it also illustrates how a transputer network can considerably speed up a computationally intensive graphics program. To demonstrate in detail the process of parallelization of an existing program, the code for the sequential ray tracing program is described and salient parts are included. Then, to illustrate just which parts of the program need to be modified to obtain a parallel solution, the relevant code fragments from the parallel version are also included.

The existing Pascal program ran on an IBM 6150 workstation (alias the IBM RT PC) under AIX, IBM's version of the UNIX operating system. The program was developed to support a computer graphics course which specializes in the display of three-dimensional objects, and it allowed scenes composed of bricks, spheres and other solids to be created and then rendered using ray tracing.

Perhaps the simplest and most general method of displaying a scene is to trace rays of light back from the viewpoint through the pixels to their origins: if a ray intersects no object in the scene the corresponding pixel is painted a background colour; if the ray intersects some object then a shading algorithm determines the colour of the object at the point of intersection, and the corresponding pixel is duly painted.

Thus each ray must, in general, be tested for intersection with each object in the scene to find the nearest object intersected. The colour of the ray reaching the viewpoint will depend upon the orientation and physical properties (colour,

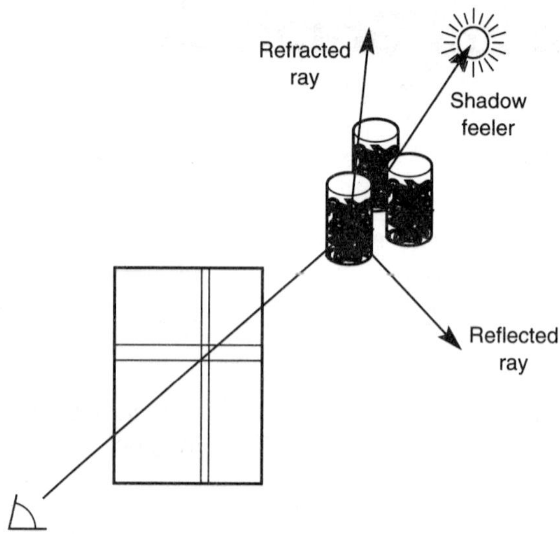

Figure 12.1

reflectivity, transparency) of the surface and the position and characteristics of each light source, direct or indirect. As each ray traced back is found to intersect a surface, we recursively trace further rays to determine the sources of the reflected and, if appropriate, the refracted rays. We also trace a 'shadow feeler' or ray from the point of intersection to each light source to establish whether there is any object in the way to cast a shadow on the surface. (See Figure 12.1.)

The cost of ray tracing arises from the large number of rays and from the number and cost of the object intersection tests for each ray. If the screen has a resolution of 1000 × 1000 pixels there will be a million primary rays, each of which may spawn several shadow feelers and also reflected and refracted rays which, in turn, may spawn further rays; furthermore, if a high quality image is required, free of jagged lines and other 'aliasing' artefacts, the 'antialiasing' technique of supersampling may be used: here several rays, perhaps four or 16, are traced through different parts of the area covered by the pixel, and the resulting colours are averaged to produce a composite colour for the pixel. Thus there may be tens or hundreds of millions of rays.

Fortunately, there are techniques to reduce this number somewhat: adaptive antialiasing means that supersampling is used only in those regions where the image is changing significantly, such as at the edges of objects; and adaptive reflection (and refraction) tracing means that new rays are recursively generated only when they can make a significant contribution to the colour of the pixel. The costs of the ray/object intersection tests can be reduced in two ways: the number of tests can be reduced by 'octree' decomposition, in which the volume spanning the scene is divided into eight octants so that only the objects in those octants on the route of the ray need be tested and any octant containing more than, say, eight objects is recursively subdivided – and of course, the testing can cease once an intersection is found; and we can surround each complex object or group of objects by a bounding box, and only if the ray intersects the box need we carry out the more expensive object intersection test. These improvements generally have the effect of making the cost of ray tracing a scene almost independent of the number of objects in the scene but several hours may still be necessary.

The ray tracing program is written as a number of text files; one contains the main program and consists mainly of 'include file' directives to the compiler; each of these included text files encapsulates all the data and routines for one logical part of the system, for instance, constructing the scene, or tracing the rays, or implementing the shading algorithm, or performing the ray intersection tests, or for handling one type of object such as a cylinder or a cone, and so on; in particular, the included file display.pas contains all the data and code peculiar to the IBM 6150 with alternative versions for other configurations such as that of the Macintosh. Thus, this was our way of writing a modular program in Pascal – it took advantage of the common but non-standard extension that allows declarations to be placed in any order. The main file, showing the principal types, variables and procedures exported by the 'modules' is as follows:

```
{ray.pas}

{This version of the ray program runs on an IBM 6150 workstation.}

program ray (input, output);

{$I display.pas}        {InitializeDisplay, DisplayPixel, DisplayLine,
                         FinalizeDisplay, etc.}
{$I maths.pas}          {vector, point, Normalize, DotProduct,
                         CrossProduct, etc.}
{$I view.pas}           {InitializeView, EyePoint, LookPoint,
                         UpVector, NextRay, etc.}
{$I lights.pas}         {InitializeLights, Background, AmbientLight,
                         MakeLight, etc.}
{$I surfaces.pas}       {BLACK, WHITE, RED, GREEN, BLUE,
                         YELLOW, SILVER, etc.}

{$I objdata.pas}        {objectptr, ObjectList, CountOfObjects, etc.}
{$I brick.pas}          {MakeBrick, etc.}
{$I sphere.pas}         {MakeSphere, etc.}
{$I cylinder.pas}       {MakeCylinder, etc.}
{$I cone.pas}           {MakeCone, etc.}
{$I objrtns.pas}        {InitializeObjects, IntObject, NrmObject,
                         SurfObject}

{$I inter.pas}          {InitializeInter, RayIntersection}
{$I shade.pas}          {InitializeShade, Shade}
```

```
{$I trace.pas}              {TraceLines}
{$I scene.pas}              {InitializeScene}

begin
        InitializeDisplay;       {sets up Width and Height in pixels of the
                                 image}
        InitializeMaths(6150);   {sets up (any) seed for random number gen-
                                 erator}
        InitializeScene;         {sets up view, lights and objects}
        InitializeInter;         {sets up the intersection data, if necessary}
        InitializeShade;         {sets up shading parameters, if necessary}
        TraceLines;              {traces rays from the eye through each pixel}
        FinalizeDisplay          {terminates the display}
end.
```

All input and output operations are restricted to the 'module' display.pas which, with much of its device dependency replaced by comments, takes the following form. The principal procedure is DisplayPixel which is called when the colour of each pixel has been computed: it stores that colour in an array lineofpixels, using dithering to give the illusion of a greater range of colours than the 256 available (8 for red × 8 for green × 4 for blue). At the end of each line of pixels, DisplayLine is called: it displays the line on the screen and, if required, writes it to a file. InitializeDisplay reads the name of the file (if any) in which the image is to be stored, and also the required width and height of the image; it sets up the colour map and clears the screen; and it initializes the 4 × 4 dither matrix:

```
{display.pas}

{This version of display is for the IBM 6150.}

{Exports: const  MAXWIDTH, MAXHEIGHT, ASPECTRATIO;
    type    widthrange, heightrange;
    var Width, Height;
    proc/fn InitializeDisplay, DisplayPixel, DisplayLine, FinalizeDisplay, Key}

const
    MAXWIDTH = 1024;        {Maximum width of image, in pixels}
    MAXHEIGHT = 1024;       {Maximum height of image, in pixels}
    ASPECTRATIO = 1.0;      {Ratio of pixel width to pixel height}
    MAXINTENSITY = 255;     {Size of the colour look up table is 256}

type
    widthrange = 1..MAXWIDTH;
    heightrange = 1..MAXHEIGHT;
    intensityrange = 0..MAXINTENSITY;
    pixelbuffer = array[widthrange] of char;
    colourtable = array[intensityrange] of integer;
    filename = packed array [1..8] of char;
```

```
var
    Width, basex : widthrange;          {image size and origin constants}
    Height, basey : heightrange;        {image size and origin constants}
    lineofpixels : pixelbuffer;         {for holding the current line of the
                                        image}
    R, G, B : colourtable;              {red, green and blue colour look up
                                        tables}
    dither : array[0..3, 0..3] of 0..15; {4 × 4 dither matrix}
    pixelfile : text;                   {for storing the generated image}
    pixelfilename : filename;           {name of the file for saving the
                                        image}
    saving : Boolean;                   {whether saving the image on file}

procedure InitializeDisplay;

    procedure ColourMap (var R, G, B : colourtable);
        var
            red, green, blue : 0..7;
            index : intensityrange;
    begin
        for red := 0 to 7 do            {3 bits for red}
            for green := 0 to 7 do      {3 bits for green}
                for blue := 0 to 3 do   {2 bits for blue}
                    begin
                        index := 32 * red + 4 * green + blue;
                        R[index] := {an IBM 6150 function of red};
                        G[index] := {an IBM 6150 function of green};
                        B[index] := {an IBM 6150 function of blue};
                    end
    end;

    var
        index : widthrange;
begin {InitializeDisplay}
    {read pixelfilename, if any, from the IBM 6150 keyboard};
    saving := pixelfilename <> '        ';
    if saving then {rewrite pixelfile using IBM 6150 convention};
    {read Width and Height of image (in pixels) from the IBM 6150
    keyboard};
    if saving then writeln(pixelfile, Width : 6, Height : 6);
    for index := 1 to MAXWIDTH do lineofpixels[index] := chr(0);
    basex := (MAXWIDTH – Width) div 2;
    basey := (MAXHEIGHT – Height) div 2;

    {initialize graphics on the IBM 6150};
    ColourMap(R, G, B);
    {set up the colour map on the IBM 6150};
    {clear the screen on the IBM 6150};
```

```
    dither[0, 0]:= 0;  dither[0, 1]:= 8;  dither[0, 2]:= 2;  dither[0, 3]:= 10;
    dither[1, 0]:= 12; dither[1, 1]:= 4;  dither[1, 2]:= 14; dither[1, 3]:= 6;
    dither[2, 0]:= 3;  dither[2, 1]:= 11; dither[2, 2]:= 1;  dither[2, 3]:= 9;
    dither[3, 0]:= 15; dither[3, 1]:= 7;  dither[3, 2]:= 13; dither[3, 3]:= 5
end; {InitializeDisplay}

procedure DisplayPixel (liney : heightrange; pixelx : widthrange; rr, gg, bb :
real);
    var
        r, g, b, i, j : intensityrange;
begin
    if rr < 1 then i := trunc(0.5 + 112 * rr) else i := 112; r := i div 16;
    if i mod 16 > dither[pixelx mod 4, liney mod 4] then r := r + 1; {0 ≤ r ≤ 7}
    if gg < 1 then i := trunc(0.5 + 112 * gg) else i := 112; g := i div 16;
    if i mod 16 > dither[pixelx mod 4, liney mod 4] then g := g + 1; {0 ≤ g ≤ 7}
    if bb < 1 then i := trunc(0.5 + 48 * bb) else i := 48; b := i div 16;
    if i mod 16 > dither[pixelx mod 4, liney mod 4] then b := b + 1; {0 ≤ b ≤ 3}
    lineofpixels[basex + pixelx] := chr(32 * r + 4 * g + b)
end;

procedure DisplayLine (liney : heightrange);
    var
        i : integer;
begin
    {output lineofpixels to line[basey + liney] of the IBM 6150 screen};
    if saving then
        begin
            for i := 1 to Width do write(pixelfile, lineofpixels[basex + i]);
            writeln(pixelfile)
        end
end;

procedure FinalizeDisplay;
begin
    if saving then close(pixelfile);
    {wait for any key on the IBM 6150 keyboard to be pressed};
    {switch out of IBM 6150 graphics mode}
end;

function Key : Boolean;
begin
    {test whether a key has been pressed on the IBM 6150 keyboard}
end;
```

The file maths.pas defines the type vector and provides several vector operations such as DotProduct and CrossProduct, since rays are represented as vectors; it also provides several other library declarations such as a random function for use with fractal and jittering techniques.

The file view.pas contains several variables such as EyePoint and FieldOfView, which are assigned values to define the position and orientation of the viewer, and it provides a procedure NextRay which, when given the coordinates of a pixel, returns a vector representing the ray from the viewer to that pixel. The file lights.pas enables one to define a Background colour for the scene and the nature of the AmbientLight illuminating the scene; using the procedure MakeLight one can define an arbitrary number of light sources, which may be positioned anywhere in the scene.

Setting up a scene entails defining not only the viewer and the lights but also, of course, the objects that make up the scene. The program shown here supports four types of object: bricks, spheres, cylinders and cones, but others may easily be added. For each we have a file, such as brick.pas, that allows us to create an object of that type (MakeBrick), determine whether and where a ray intersects the object (IntBrick) and, if it does, what the normal to the object at the point of intersection is (NrmBrick); we can also interrogate the object to find the colour of its surface (SurfBrick). A library of surface types is provided by the file surfaces.pas. The files enveloping the several object files define the data structures required for representing the objects (in objdata.pas) and some generic routines: IntObject, NrmObject and SurfObject (in objrtns.pas).

The file inter.pas provides a Boolean function RayIntersection, which uses IntObject to check which object, if any, a particular ray first intersects. Depending upon the version used, this may entail an exhaustive check of all the objects in the scene, or it may use recursive octree decomposition to limit the search to objects near the path of the ray. Given that a ray intersects an object, we can compute the colour at the point of intersection by invoking the procedure Shade of the file shade.pas. This sums the amount of ambient light reflected, the amounts of light reflected diffusely and specularly from all the light sources that illuminate the point of intersection and, perhaps, the reflections (and refractions if the medium is transparent) of other objects.

The procedure TraceLines in file trace.pas is responsible for the ray tracing, using the routines NextRay (from view.pas), RayIntersection (from inter.pas), Shade (from shade.pas) and (from display.pas) DisplayPixel, DisplayLine and, to enable the ray tracing to be interrupted, Key. More elaborate versions than the one shown here are normally used, incorporating, for example, adaptive jittered supersampling to overcome aliasing effects such as jagged lines at the edges of objects:

```
{trace.pas}

{This version of trace generates one ray per pixel and performs no antialias-
ing.}

{Export:    procedure TraceLines}

procedure TraceLines;
```

```
    label
        999;

    var
        pixelx : widthrange;
        liney : heightrange;
        ray : vector;
        colour : colourtype;

    procedure TraceRay (pixelx, liney : real;  var colour : colourtype);
        var
            ray, normal : vector;
            obj : objectptr;
            hitpoint : point;
            surf : surfacetype;
    begin
        NextRay(pixelx, liney, ray);
        if RayIntersection(EyePoint, ray, obj, hitpoint, normal, surf)
        then Shade(ray, obj, hitpoint, normal, surf, colour)
        else colour := Background
    end;

begin  {TraceLines}
    for liney := 1 to Height do
        begin
            for pixelx := 1 to Width do
            begin
                if Key then goto 999;
                {compute r, g, b values for pixelx on liney}
                TraceRay(pixelx, liney, colour);
                with colour do DisplayPixel(liney, pixelx, r, g, b)
            end;
            DisplayLine(liney)
        end;
999 :
end;  {TraceLines}
```

The final file, scene.pas, has a single procedure, InitializeScene, in which the view is defined, the lights are created and the objects constituting the scene are defined and positioned. All these files are only a page or two long, except possibly for display.pas, which is usually the longest.

12.2 The sequential program

Ray tracing on the IBM 6150 was expensive: perhaps the best part of a day for a 1024 × 1024 image of a complex scene. It was at this point that transputers came to our aid. We moved the program from a UNIX workstation to a PC with a board containing a single T800 transputer – and promptly doubled the speed of the program despite using very much cheaper hardware. What were the problems of porting the sequential program from one machine to another? Two minor problems arose from the common practice of using non-standard extensions to Pascal for including source files and for rewinding data files. The other problem concerned the output of the ray tracer: should the image be written to a file or displayed on a screen?

The notation for including text files in 3L Pascal is $include 'display.pas' rather than {$I display.pas} so 15 such lines had to be changed:

```
{ray.pas}

{This version of the ray program runs on a single transputer.}

program ray (input, output);

$include 'display.pas'      {InitializeDisplay, DisplayPixel, DisplayLine,
                            FinalizeDisplay, etc.}
$include 'maths.pas'        {vector, point, Normalize, DotProduct,
                            CrossProduct, etc.}
$include 'view.pas'         {InitializeView, EyePoint, LookPoint, UpVector,
                             NextRay, etc.}
$include 'lights.pas'       {InitializeLights, Background, AmbientLight,
                            MakeLight, etc.}
$include 'surfaces.pas'     {BLACK, WHITE, RED, GREEN, BLUE,
                            YELLOW, SILVER, etc.}

$include 'objdata.pas'      {objectptr, ObjectList, CountOfObjects, etc.}
$include 'brick.pas'        {MakeBrick, etc.}
$include 'sphere.pas'       {MakeSphere, etc.}
$include 'cylinder.pas'     {MakeCylinder, etc.}
$include 'cone.pas'         {MakeCone, etc.}
$include 'objrtns.pas'      {InitializeObjects, IntObject, NrmObject,
                            SurfObject}

$include 'inter.pas'         {InitializeInter, RayIntersection}
$include 'shade.pas'        {InitializeShade, Shade}
$include 'trace.pas'         {TraceLines}
$include 'scene.pas'         {InitializeScene}

begin
   InitializeDisplay;       {sets up Width and Height in pixels of the
                            image}
   InitializeMaths(6150);   {sets up (any) seed for random number generator}
   InitializeScene;         {sets up view, lights and objects}
   InitializeInter;         {sets up the intersection data, if necessary}
   InitializeShade;         {sets up shading parameters, if necessary}
```

```
    TraceLines;             {traces rays from the eye through each pixel}
    FinalizeDisplay         {terminates the display}
end.
```

The IBM 6150 file rewinding statement rewrite(pixelfile, 'NAME=abcdefgh.pic') must be rendered as rewrite(pixelfile, 'abcdefgh.pic') for the transputer. On the IBM 6150 the ray tracer always displays the image on the screen and, if the user supplies the name of a file, the image is written to that file, pixel by pixel, together with a header consisting of the width and height of the image. As our PCs did not have high resolution screens we chose to make the transputer version of display.pas just write the image to disk. We used separate programs to display filed images on the IBM 6150 screen or on a high resolution screen attached to the transputer network; these programs set up their own colour maps. The Boolean function Key is used in the IBM 6150 version for two purposes: one is to allow the user to stop the ray tracer prematurely; the other is to hold the finished image on the screen until the user has had time to see it. On the transputer system the latter is not required, and the former can be accomplished by interrupting the server program on the PC. The result of all this is that the transputer version of display.pas is very much simpler and shorter than the IBM 6150 version:

```
{display.pas}

{This version of display is for a single transputer.}

{Exports: const  MAXWIDTH, MAXHEIGHT, ASPECTRATIO;
    type    widthrange, heightrange;
    var Width, Height;
    proc/fn InitializeDisplay, DisplayPixel, DisplayLine, FinalizeDisplay, Key}

const
    MAXWIDTH = 1024;       {Maximum width of image, in pixels}
    MAXHEIGHT = 1024;      {Maximum height of image, in pixels}
    ASPECTRATIO = 1.0;     {Ratio of pixel width to pixel height}
    MAXINTENSITY = 255;    {Size of the colour look up table is 256}

type
    widthrange = 1..MAXWIDTH;
    heightrange = 1..MAXHEIGHT;
    intensityrange = 0..MAXINTENSITY;
    filename = packed array [1..8] of char;

var
    Width : widthrange;                  {image size constant}
    Height : heightrange;                {image size constant}
    dither : array[0..3, 0..3] of 0..15; {4 × 4 dither matrix}
    pixelfile : text;                    {for storing the generated image}
    pixelfilename : filename;            {name of the file for saving the image}
```

```
procedure InitializeDisplay;
begin
    {read pixelfilename from the PC keyboard};
    {rewrite pixelfile using 3L convention};
    {read Width and Height of image (in pixels) from the PC keyboard};
    writeln(pixelfile, Width : 6, Height : 6);

    dither[0, 0]:= 0;  dither[0, 1]:= 8;  dither[0, 2]:= 2;  dither[0, 3]:= 10;
    dither[1, 0]:= 12; dither[1, 1]:= 4;  dither[1, 2]:= 14; dither[1, 3]:= 6;
    dither[2, 0]:= 3;  dither[2, 1]:= 11; dither[2, 2]:= 1;  dither[2, 3]:= 9;
    dither[3, 0]:= 15; dither[3, 1]:= 7;  dither[3, 2]:= 13; dither[3, 3]:= 5
end;

procedure DisplayPixel (liney : heightrange;  pixelx : widthrange;  rr, gg, bb :
real);
    var
        r, g, b, i, j : intensityrange;
begin
    if rr < 1 then i := trunc(0.5 + 112 * rr) else i := 112;  r := i div 16;
    if i mod 16 > dither[pixelx mod 4, liney mod 4] then r := r + 1;    {0≤r≤7}
    if gg < 1 then 1 := trunc(0.5 + 112 * gg) else i := 112; g := i div 16;
    if i mod 16 > dither[pixelx mod 4, liney mod 4] then g := g + 1;    {0≤g≤7}
    if bb < 1 then i := trunc(0.5 + 48 * bb) else i := 48; b := i div 16;
    if i mod 16 > dither[pixelx mod 4, liney mod 4] then b := b + 1;    {0≤b≤3}
    write(pixelfile, chr(32 * r + 4 * g + b))
end;

procedure DisplayLine (liney : heightrange);
begin  writeln(pixelfile)  end;

procedure FinalizeDisplay;
begin  close(pixelfile)  end;

function Key : Boolean;
begin  Key:= false  end;
```

The other 14 'modules' are unaffected by the conversion. The resulting transputer program was run with a test scene of 1015 spheres arranged in a pyramid: generating a 512 × 512 image took 151 minutes on a T800 transputer, whereas it had taken 323 minutes on an IBM 6150. However, two and a half hours is still a long time to wait for an image and, for a further improvement, we had to go parallel.

12.3 The parallel program

Fortunately, ray tracing lends itself perfectly to parallelization. Subimages can be farmed out to all the processors, each of which runs a copy of the ray tracer, with a small master program to distribute the work and collect the results. With, say, four processors, one could give each a quarter of the image, but there is a possibility that the load would be unevenly spread; so the method is to farm out units of work that are smaller, but not so small that the overheads dominate the useful computation. We took a row of the image as the unit. As we moved from one transputer to four the time to trace the pyramid of 1015 spheres for a 512 × 512 image changed from 151 minutes to 44 minutes: on the network of four transputers, 86% of the time was being used profitably and 14% was wasted. Not all that 14% can be regained: whatever the size of the unit the data will have to be transmitted over the network, and some duplication of work is inevitable as all the processors engage in the same initialization such as setting up the scene to be rendered. Perhaps some improvement could be made in the unit size, but a line at a time is convenient and no amount of fine tuning will result in great savings. An argument can be made for using a square or rectangular subimage as the unit rather than a row of the image: adaptive antialiasing requires that supersampling of a pixel should be done if that pixel differs significantly in colour from any of its immediate neighbours (the purpose of this is to smooth the 'jaggies' that can mar the edge of an object).

3L's processor farm proved almost ideal for our application: a **master** task farms out lines to be rendered to several identical **worker** tasks which generate the pixels and return them to the master. The 3L software allows the user to program this without the need to know, even at runtime, how many worker tasks there are and where they are located. The flood-fill configurer will automatically place a worker on each transputer in the network and the master on the root transputer, and it will look after the runtime scheduling of workers and the routing of messages between the master and the workers by placing a small routing task on each transputer.

Each worker uses the library routine net_receive to receive a work packet and it sends its results back using net_send. It cannot communicate with the host (using operations such as read or writeln), or with other transputer tasks using operations such as 3L's chan_out_message. The master task generates the packets of work and sends them to the anonymous workers using net_send; it also receives the results using net_receive. The way in which the master multiplexes itself among these operations is by using multiple 'threads': it uses the operation thread_create to run a procedure as a subtask in quasiparallel with itself. The master takes the following form:

```
{raymstr.pas}

{This version of the ray program runs on the root transputer in a processor
farm.}

program raymstr (input, output);

$include 'net.inc'
$include 'thread.inc'
$include 'display.pas'
```

```
    const
        workspacesize = 1000;                       {approx. bytes required
                                                    by thread}
    type
        byte = 0..255;
    var
        workspace: array [1..workspacesize] of byte; {workspace required by
                                                    thread}
        threadok: Boolean;                          {result of creating
                                                    thread}

procedure SEND;
    type
        packet =  record
                      lineno: heightrange;
                      ASPECTRATIO: real;
                      Width: widthrange;
                      Height: heightrange
                  end;
    var
        P: packet;
        liney: heightrange;
begin
    P.ASPECTRATIO:= ASPECTRATIO;
    P.Width:= Width;
    P.Height:= Height;
    for liney:= 1 to Height do
    begin  P.lineno:= liney;  Net_Send(Size(P), P, true)  end
end;

procedure RECEIVE;
    var
        liney: heightrange;
        lineno: heightrange;
        length: integer;
        complete: Boolean;
begin
    for liney:= 1 to Height do
    begin
        length:= Net_Receive(lineno, complete);
        length:= Net_Receive(pixelarray[lineno], complete)
    end
end;

begin  {raymstr}
    InitializeDisplay;
    threadok:= Thread_Create(SEND, workspace, 0);
    RECEIVE;
    FinalizeDisplay
end.  {raymstr}
```

where:

```
{display.pas}

{This version of display is for the master in a processor farm.}

{Exports: const   MAXWIDTH, MAXHEIGHT, ASPECTRATIO;
    type    widthrange, heightrange;
    var Width, Height, pixelarray;
    proc/fn InitializeDisplay, FinalizeDisplay, Key}

const
    MAXWIDTH = 1024;       {Maximum width of image, in pixels}
    MAXHEIGHT = 1024;      {Maximum height of image, in pixels}
    ASPECTRATIO = 1.0;     {Ratio of pixel width to pixel height}

type
    widthrange = 1..MAXWIDTH;
    heightrange = 1..MAXHEIGHT;
    pixelbuffer = array [widthrange] of char;
    filename = packed array [1..8] of char;

var
    Width : widthrange;                          {image size constant}
    Height : heightrange;                        {image size constant}
    pixelarray : array [heightrange] of pixelbuffer; {for storing the generated
                                                  image}
    pixelfile : text;                            {for storing the generated
                                                  image}
    pixelfilename : filename;                    {name of file for saving the
                                                  image}

procedure InitializeDisplay;
begin
    {read pixelfilename from the PC keyboard};
    {rewrite pixelfile using 3L convention};
    {read Width and Height of image (in pixels) from the PC keyboard};
    writeln(pixelfile, Width : 6, Height : 6)
end;

procedure DisplayPixel (liney : heightrange;  pixelx : widthrange;  rr, gg, bb :
real);
begin  {not used in this version}  end;

procedure DisplayLine (liney : heightrange);
begin  {not used in this version}  end;

procedure FinalizeDisplay;
    var
```

```
        pixelx : widthrange;
        liney : heightrange;
begin
    for liney:= 1 to Height do
    begin
        for pixelx:= 1 to Width do write(pixelfile, pixelarray[liney, pixelx]);
        writeln(pixelfile)
    end
end;

function Key : Boolean;
begin  Key:= false  end;
```

In the master the subtask SEND runs concurrently with the main task, which is an invocation of RECEIVE; true concurrency is not possible because both 'threads' run on the same transputer. As packets may take different times to process, perhaps because of the data they contain or because they are processed on transputers of different speeds, it is necessary to transmit the line number with the result packet so that the lines can be assembled in the right order. As each packet is limited to 1024 bytes and the line may be as large as that, it is necessary to transmit the line number as a separate packet. Note that the main thread of the task must not terminate until all the other threads have finished their useful work: this is the case here where all the lines that are sent by SEND must subsequently be received back by RECEIVE. Although parameters can be passed to SEND when the thread is created, it is often simpler to transmit them over an internal channel or, in this case, to let SEND read the values of the global variables itself – this can be done without danger of mutual exclusion problems as the values are constant during the currency of the thread. There was a problem in transmitting the parameters to the workers: each time we asked for a line to be traced we had to transmit the initialization parameters, too; this was because there was no way of initializing workers other than sending them normal work packets.

As we use the flood-fill configurer the commendably brief configuration file takes the following form:

```
! ray.cfg
task  master  file=raymstr  data=1100k  ! large enough to hold a 1024x1024
                                          image
task  worker  file=raywrkr
```

The worker tasks have the following form:

```
{raywrkr.pas}

{This version of the ray program runs on every transputer in a processor
farm.}
```

```
module raywrkr;

$include 'net.inc'

$include 'display.pas'    {InitializeDisplay, DisplayPixel, DisplayLine,
                           FinalizeDisplay, etc.}
$include 'maths.pas'      {vector, point, Normalize, DotProduct,
                           CrossProduct, etc.}
$include 'view.pas'       {InitializeView, EyePoint, LookPoint, UpVector,
                           NextRay, etc.}
$include 'lights.pas'     {InitializeLights, Background, AmbientLight,
                           MakeLight, etc.}
$include 'surfaces.pas'   {BLACK, WHITE, RED, GREEN, BLUE, YELLOW,
                           SILVER, etc.}

$include 'objdata.pas'    {objectptr, ObjectList, CountOfObjects, etc.}
$include 'brick.pas'      {MakeBrick, etc.}
$include 'sphere.pas'     {MakeSphere, etc.}
$include 'cylinder.pas'   {MakeCylinder, etc.}
$include 'cone.pas'       {MakeCone, etc.}
$include 'objrtns.pas'    {InitializeObjects, IntObject, NrmObject, SurfObject}

$include 'inter.pas'      {InitializeInter, RayIntersection}
$include 'shade.pas'      {InitializeShade, Shade}
$include 'trace.pas'      {TraceLines}
$include 'scene.pas'      {InitializeScene}

begin
   InitializeDisplay;         {sets up Width and Height in pixels of the image}
   InitializeMaths(6150);     {sets up (any) seed for random number generator}
   InitializeScene;           {sets up view, lights and objects}
   InitializeInter;           {sets up the intersection data, if necessary}
   InitializeShade;           {sets up shading parameters, if necessary}
   TraceLines;                {traces rays from the eye through each pixel}
   FinalizeDisplay            {terminates the display}
end.
```

3L Pascal has the curious rule that programs that use the standalone runtime library must use the keyword module rather than program and must render global variables as, say:

```
static x : xtype;
static y : ytype;
static z : ztype;
```

instead of:

```
var
        x : xtype;
        y : ytype;
        z : ztype;
```

Thus, global variables in all the included files have to be redeclared; this is mildly tedious but not too onerous, as most have no global variables and the rest (apart perhaps from display.pas) have very few.

The version of display.pas used in the worker is as follows:

```
{display.pas}

{This version of display is for the worker in a processor farm.}

{Exports:  const  MAXWIDTH, MAXHEIGHT, ASPECTRATIO;
    type   widthrange, heightrange;
    var Width, Height, liney;
    proc/fn InitializeDisplay, DisplayPixel, DisplayLine, Key}

const
    MAXWIDTH = 1024;          {Maximum width of image, in pixels}
    MAXHEIGHT = 1024;         {Maximum height of image, in pixels}
    MAXINTENSITY = 255;       {Size of the colour look up table is 256}

type
    widthrange = 1..MAXWIDTH;
    heightrange = 1..MAXHEIGHT;
    intensityrange = 0..MAXINTENSITY;
    pixelbuffer = array[widthrange] of char;
    colourtable = array[intensityrange] of integer;
    packet = record
                lineno: heightrange;
                ASPECTRATIO: real;
                Width: widthrange;
                Height: heightrange
                  end;

static  ASPECTRATIO : real;                {ratio of pixel width to pixel height}
static  Width : widthrange;                {image size constant}
static  Height : heightrange;              {image size constant}
static  lineofpixels : pixelbuffer;        {for holding the current line of the
                                            image}
static  dither : array[0..3, 0..3] of 0..15;  {4 × 4 dither matrix}
static  pixelfile : text;                  {for storing the generated image}
static  pixelfilename : filename;          {name of file for saving the image}
static  liney : heightrange;               {the number of the current line to
                                            be traced}
```

```
static  length : integer;            {length of packet received over
                                      network}
static  complete : Boolean;          {denotes whether last packet of a
                                      message}
static  P : packet;                  {packet transmitted over network}

procedure InitializeDisplay;

begin
   length:= Net_Receive(P, complete);
   liney:= P.lineno;
   ASPECTRATIO:= P.ASPECTRATIO;
   Width:= P.Width;
   Height:= P.Height;

   dither[0, 0]:=  0;  dither[0, 1]:=  8;  dither[0, 2]:=  2;  dither[0, 3]:= 10;
   dither[1, 0]:= 12;  dither[1, 1]:=  4;  dither[1, 2]:= 14;  dither[1, 3]:=  6;
   dither[2, 0]:=  3;  dither[2, 1]:= 11;  dither[2, 2]:=  1;  dither[2, 3]:=  9;
   dither[3, 0]:= 15;  dither[3, 1]:=  7;  dither[3, 2]:= 13;  dither[3, 3]:=  5
end;

procedure DisplayPixel (liney : heightrange;  pixelx : widthrange;  rr, gg, bb :
real);
   var
      r, g, b, i, j : intensityrange;
begin
   if rr < 1 then i := trunc(0.5 + 112 * rr) else i := 112;  r := i div 16;
   if i mod 16 > dither[pixelx mod 4, liney mod 4] then r := r + 1;    {0 ≤ r ≤ 7}
   if gg < 1 then i := trunc(0.5 + 112 * gg) else i := 112;  g := i div 16;
   if i mod 16 > dither[pixelx mod 4, liney mod 4] then g := g + 1;    {0 ≤ g ≤ 7}
   if bb < 1 then i := trunc(0.5 +  48 * bb) else i := 48;   b := i div 16;
   if i mod 16 > dither[pixelx mod 4, liney mod 4] then b := b + 1;    {0 ≤ b ≤ 3}
   lineofpixels[pixelx]:= chr(32 * r + 4 * g + b)
end;

procedure DisplayLine (liney : heightrange);
begin
   Net_Send(Size(liney), liney, false);
   Net_Send(Width * Size(char), lineofpixels, true);
   length:= Net_Receive(P, complete);
   liney:= P.lineno
end;

procedure FinalizeDisplay;
begin  {not used in this version}  end;

function Key : Boolean;
begin  Key:= false  end;
```

Apart from declaring the remaining global variables in the other included files as static (as above in display.pas), the only change that needs to be made is to the file trace.pas. The declaration of liney is removed as the value in display.pas is used, and the for loop is replaced by while true do as new values of liney are supplied by the master whenever DisplayLine is invoked. More elaborate versions of trace.pas are normally used; the following version, for use in a processor farm, illustrates the use of jittered supersampling for antialiasing:

```
{trace.pas}

{This farm version of trace generates four jittered rays per pixel for antialias-
ing.}

{Export:   procedure TraceLines}

procedure TraceLines;

    label
        999;

    static  pixelx : widthrange;                   {number of pixel on line}
    static  liney : heightrange;                   {number of line on image}
    static  ray : vector;                          {the ray to be traced}
    static  colour, c1, c2, c3, c4 : colourtype;   {colours from tracing rays}

    procedure TraceRay (pixelx, liney : real;  var colour : colourtype);
        var
            ray, normal : vector;
            obj : objectptr;
            hitpoint : point;
            surf : surfacetype;
    begin
        NextRay(pixelx, liney, ray);
        if RayIntersection(EyePoint, ray, obj, hitpoint, normal, surf)
        then Shade(ray, obj, hitpoint, normal, surf, colour)
        else colour := Background
    end;

begin {TraceLines}
    while true do
        begin
            for pixelx := 1 to Width do
            begin
                if Key then goto 999;
                {compute r, g, b values for pixelx on liney}
                TraceRay(pixelx–random/4, liney–random/4, c1);
                TraceRay(pixelx–random/4, liney+random/4, c2);
                TraceRay(pixelx+random/4, liney–random/4, c3);
```

```
                TraceRay(pixelx+random/4, liney+random/4, c4);
                colour.r:= (c1.r + c2.r + c3.r + c4.r)/4;
                colour.g:= (c1.g + c2.g + c3.g + c4.g)/4;
                colour.b:= (c1.b + c2.b + c3.b + c4.b)/4;
                with colour do DisplayPixel(liney, pixelx, r, g, b)
            end;
            DisplayLine(liney)
        end;
999:
end; {TraceLines}
```

12.4 In conclusion

Since the parallel version of the ray tracer was developed, 3L has expanded the facilities for operating a processor farm. One useful addition has been the provision of an operation net_broadcast(length, message) to enable the master to initialize all the workers, each of which uses the net_receive operation to read the message, packet by packet; this would have been useful for passing the values of ASPECTRATIO, Width and Height and for changing the viewing parameters or the scene.

Another improvement has been to permit mixed configurations in which subsidiary networks (flood-filled or otherwise) are attached to the edges of the main network; thus certain transputers on the periphery of the main network act as hosts to the subsidiary networks. The configuration file can also cater for processors other than transputers. This would have enabled us to reserve one of the transputers to drive the high resolution screen while the remainder formed the processor farm.

These amendments have alleviated two of the most pressing problems we encountered:

- there was no clean way to initialize the workers, and
- there was no clean way to attach extra processes to a processor farm.

The other, mainly minor, problems we met were:

- the use of non-standard extensions for 'include file' directives and for rewrite statements had to be modified,
- differences in input and output necessitated fresh versions of display.pas,
- a master task with its own version of display.pas had to be written for the processor farm,
- the worker task required minor cosmetic changes: module and static,
- the worker's trace.pas required a minor change regarding the numbers of the lines to be traced,

- the compiler occasionally generates erroneous calls to a library routine PASCAL_CLOSE; the fault apparently arises at random, and has never been satisfactorily resolved.

The verdict? A worthwhile improvement in performance for little effort. We have also produced other parallel versions of the ray tracer: for example, one which produces both left and right eye views for stereoscopic images, one which runs on a Parsys SuperNode using the Tiny message passing primitives instead of the net operations, and one which runs on a network of DEC 3100 workstations. More time is spent becoming acquainted with the peculiarities of each machine and its software than on adapting the ray tracer.

Acknowledgements

This ray tracer was originally based on an excellent one written in C by Roman Kuchkuda, and described in 'An Introduction to Ray Tracing' (pp. 1039–1059, NATO Advanced Study Institute Series, volume F40, *Theoretical Foundations of Computer Graphics and CAD*, R. A. Earnshaw, ed., 1988, Springer-Verlag). That lucid exposition inspired me to implement a complete ray tracer to illustrate my lectures on the subject. Complete? No ray tracer is ever complete; there is always something else to add: new optical phenomena, new types of object, new parallel architectures.

This Pascal program has been modified many times for many purposes by many students; in particular, my thanks go to Sam Burns who implemented the first transputer version and to Adele Crothers, Pall Gudmundson, Neil Irwin, Zoltan Juhasz, Eddie Killen and Surendra Shukla who developed various parallel versions over the years; few programs have been more fun to implement.

Index